Operational Risk Management

Statistics in Practice

Statistics in Practice is an important international series of texts which provide detailed coverage of statistical concepts, methods and worked case studies in specific fields of investigation and study.

With sound motivation and many worked practical examples, the books show in down-to-earth terms how to select and use an appropriate range of statistical techniques in a particular practical field within each title's special topic area.

The books provide statistical support for professionals and research workers across a range of employment fields and research environments. Subject areas covered include medicine and pharmaceutics; industry, finance and commerce; public services; the earth and environmental sciences, and so on.

The books also provide support to students studying statistical courses applied to the above areas. The demand for graduates to be equipped for the work environment has led to such courses becoming increasingly prevalent at universities and colleges.

It is our aim to present judiciously chosen and well-written workbooks to meet everyday practical needs. Feedback of views from readers will be most valuable to monitor the success of this aim.

A complete list of titles in the series appears at the end of this volume.

Operational Risk Management

A Practical Approach to Intelligent Data Analysis

Edited by

Ron S. Kenett

*KPA Ltd, Raanana, Israel; University of Turin, Italy; and
NYU-Poly, Center for Risk Engineering, New York, USA*

Yossi Raanan

*KPA Ltd, Raanana, Israel; and
College of Management, Academic Studies, Rishon Lezion, Israel*

A John Wiley and Sons, Ltd., Publication

Library of Congress Cataloging-in-Publication Data

Operational risk management : a practical approach to intelligent data analysis / edited by Ron S. Kenett, Yossi Raanan.
    p. cm.
  Includes bibliographical references and index.
  ISBN 978-0-470-74748-3 (cloth)
  1. Risk management. 2. Quality control. 3. Information technology–Quality control. 4. Process control. I. Kenett, Ron. II. Raanan, Yossi.
  HD61.O66 2010
  658.15′5–dc22

                                                         2010024537

A catalogue record for this book is available from the British Library.

Print ISBN: 978-0-470-74748-3
ePDF ISBN: 978-0-470-97256-4
oBook ISBN: 978-0-470-97257-1

Set in 10/12pt Times by Laserwords Private Limited, Chennai, India
Printed and bound in Singapore by Markono Print Media Pte Ltd.

In memory of Roberto Gagliardi

Contents

PART II   DATA FOR OPERATIONAL RISK MANAGEMENT AND ITS HANDLING   39

Foreword

The recognition from the Basel Committee of Banking Supervisors of operational risks as a separate risk management discipline has promoted in the past years intense and fruitful discussions, both inside and outside the banking and financial sectors, on how operational risks can be managed, assessed and prevented, or at least mitigated.

However, for several reasons, including the fact that operational risks appear at the same time multifaceted and of a somewhat indefinite shape, inadequate attention has been given so far to what operational risks really are, and to how they can be correctly identified and captured.

Indeed, the first objective of a risk management programme is to identify clearly the playing field to where investments and resources should be directed. This is even more important for operational risk management, since its scope crosses all industry sectors and all types of firms and the fact that it essentially originates from those variables that constitute the heart of any organization: people, processes and systems.

This book attempts to give an appropriate level of attention to this significant topic by using an interdisciplinary, integrated and innovative approach.

The methodologies and techniques outlined here, reading 'behind and beyond' operational risks, aim to move forward in the interpretation of this type of risk and of the different ways it can reveal. The objective of capturing knowledge on operational risk, rather than just information, is crucial for the building of sound processes for its management, assessment and prevention or mitigation.

Another noteworthy feature of this work is the effort – pursued by providing practical examples of implementation of an operational risk framework (or part of it) in different industry sectors – to demonstrate how concepts, techniques and methodologies developed in a specific field, for the handling of operational risks, can be adopted in (or adapted to) other industrial domains. If considered all together, these aspects can significantly contribute to make this discipline evolve towards high, sustainable and convergent standards and, above all, to change its nature from (a bit less) 'art' to (a bit more) 'science', which, in the end, is the ultimate objective that all operational risk managers are trying to achieve.

Marco Moscadelli
Bank of Italy and Committee of European Banking Supervisors

Preface

This book is a result of the MUSING (MUlti-industry, Semantic-based, next generation business INtelligence) project collaborative effort, an R&D venture co-funded by the European Commission under the FP6 Information Society Technology Programme. The project covered a period of four years, witnessing many dramatic events, including the Wall Street crash in September 2008. It was designed to be driven by customer requirements in three main areas of application: operational risk management, financial risk management and internationalization. The idea was to develop innovative solutions to customer requirements in these three domains, with partners that are leaders in their fields. The MUSING partners combined expertise and experience in risk management, information extraction and natural language processing, with ontology engineering, data mining and statistical modelling. The focus in this book is operational risk management. The customers we had in mind are financial institutions implementing Basel II regulations, industrial companies developing, manufacturing and delivering products and services, health care organizations and others with exposure to operational risks with potential harmful effects and economic impact.

The key visionaries behind the project were Roberto Gagliardi and Paolo Lombardi. At the inaugural meeting, on 5 April 2006 in Pisa, Italy, they presented a slide with the project components organized in the shape of a sailboat. The basic research and integration partners were responsible for the keel and cockpit. The sails, pushing the boat forward, were the three application areas. This strategy, where customers and users motivate researchers to develop innovative solutions based on state-of-the-art technologies, is what MUSING was about.

Unfortunately Roberto passed away as the boat started sailing, but the MUSING vision kept the project on track.

Operational risk management is a complicated topic. Its precise definition is elusive and its 'boundary of meaning' has evolved over time. From a classification of residual risks, that is risks not identified as financial risks or market risks, it has become a key area with specific methods, dedicated technology and dedicated indicators and scoring systems. This book covers many of the state-of-the-art techniques in this area, with many implementation examples from real life from MUSING. Some chapters are more mathematical than others, some are fully descriptive. In designing the book we wanted to balance the various disciplines involved in setting up an infrastructure for modern operational risk management.

The creative synergy between experts in various disciplines has made MUSING a unique project. We hope the book will convey this message to the readers. Not all the authors who contributed to the book were part of the MUSING project. All chapters, however, present the latest advances in operational risk management by a combination of novel methods in the context of real problems, as envisaged in the MUSING project. As such, we believe that the book provides a solid foundation and challenging directions for operational risk management.

In preparing an edited volume, it is natural for many people to be involved. As editors, it was a pleasure and a privilege to work with the authors of the 14 chapters. These authors were also kind enough to serve as internal reviewers. The typos and mistakes that sneaked in remain, however, our responsibility. We want to thank the authors who dedicated their time and talent to write these chapters and all our colleagues in the MUSING project who helped develop this knowledge. Special thanks are due to the project coordinators Marcus Spies and Thierry Declerk, the Metaware team, the project reviewers, Professors Vadim Ermolayev, Mark Lycett, Aljosa Pasic and the project officer, Francesco Barbato – they all contributed significantly to the success of MUSING. The help of Drs Emil Bashkansky and Paolo Lombardi in reviewing the chapters is also gratefully acknowledged. Finally, we would like to thank Dr Ilaria Meliconi, Heather Kay and Richard Davies from John Wiley & Sons, Ltd for their help, directions and patience.

Ron S. Kenett and Yossi Raanan

Introduction

Operational risk management is becoming a key competency for organizations in all industries. Financial institutions, regulated by the Basel II Accord, need to address it systematically since their level of implementation affects their capital requirements, one of their major operational expenses. Health organizations have been tackling this challenge for many years. The Institute of Medicine reported in 2000 that 44 000–98 000 patients die each year in the United States as a result of medication errors, surgical errors and missed diagnoses, at an estimated cost to the US economy of $17–29 billion. Operational risks affect large organizations as well as small and medium-sized enterprises (SMEs) in virtually all industries, from the oil and gas industry to hospitals, from education to public services.

This multi-author book is about tracking and managing operational risks using state-of-the-art technology that combines the analysis of qualitative, semantic, unstructured data with quantitative data. The examples used are mostly from information technology but the approach is general. As such, the book provides knowledge and methods that can have a substantial impact on the economy and quality of life.

The book has four main parts. Part I is an introduction to operational risk management, Part II deals with data for operational risk management and its handling, Part III covers operational risk analytics and Part IV concludes the book with several applications and a discussion on how operational risk management integrates with other disciplines. The 14 chapters and layout of the book are listed below with short descriptions.

Part I: Introduction to Operational Risk Management

This first part of the book is introductory with a review of modern risk management in general and a presentation of specific aspects of operational risk management issues.

Chapter 1: *Risk management: a general view* (R. Kenett, R. Pike and Y. Raanan)
This chapter introduces the concepts of risk management and positions operational risk management within the overall risk management landscape. The topics covered include definitions of risks, aspects of information quality and a discussion of state-of-the-art enterprise risk management. The organizations the

authors have in mind are financial institutions implementing Basel II regulations, industrial companies developing, manufacturing and delivering products and services, health care services and others with exposure to risks with potential harmful effects. The chapter is meant to be a general introduction to risk management and a context-setting background for the 13 other chapters of the book.

Chapter 2:   *Operational risk management: an overview* (Y. Raanan, R. Kenett and R. Pike)

This chapter introduces the general concepts of operational risk management in the context of the overall risk management landscape. Section 2.2 provides a definition of operational risk management, Section 2.3 covers the key techniques of this important topic, Section 2.4 discusses statistical models and Section 2.5 covers several measurement techniques for assessing operational risks. The final section summarizes the chapter and provides a roadmap for the book.

Part II: Data for Operational Risk Management and its Handling

Operational risk management relies on diverse data sources, and the handling and management of this data requires novel approaches, methods and implementations. This part is devoted to these concepts and their practical applications. The applications are based on case studies that provide practical, real examples for the practitioners of operational risk management.

Chapter 3:   *Ontology-based modelling and reasoning in operational risks* (C. Leibold, H.-U. Krieger and M. Spies)

This chapter discusses the design principles of operational risk ontologies for handling semantic unstructured data in operational risk management (OpR). In particular, the chapter highlights the contribution of ontology modelling to different levels of abstraction in OpR. Realistic examples from the MUSING project and application-domain-specific ontologies are provided. A picture is drawn of axiomatic guidelines that provide a foundation for the ontological framework and refers to relevant reporting and compliance standards and generally agreed best practices.

Chapter 4:   *Semantic analysis of textual input* (H. Saggion, T. Declerck and K. Bontcheva)

Information extraction is the process of extracting from text specific facts in a given target domain. The chapter gives an overview of the field covering components involved in the development and evaluation of an information extraction system such as parts of speech tagging or named entity recognition. The chapter introduces available tools such as the GATE system and illustrates rule-based approaches to information extraction. An illustration of information extraction in the context of the MUSING project is presented.

Chapter 5:   *A case study of ETL for operational risks* (V. Grossi and A. Romei)

Integrating both internal and external input sources, filtering them according to rules and finally merging the relevant data are all critical aspects of business analysis and risk assessment. This is especially critical when internal loss data is not sufficient for effective calculation of risk indicators. The class of tools responsible for these tasks is known as *extract, transform and load (ETL)*. The chapter reviews state-of-the-art techniques in ETL and describes an application of a typical ETL process in the analysis of causes of operational risk failures. In particular, it presents a case study in information technology operational risks in the context of a telecommunication network, highlighting the data sources, the problems encountered during the data merging and finally the solution proposed and implemented by means of ETL tools.

Chapter 6:   *Risk-based testing of web services* (X. Bai and R. Kenett)

A fundamental strategy for mitigating operational risks in web services and software systems in general is testing. Exhaustive testing of web services is usually impossible due to unavailable source code, diversified user requirements and the large number of possible service combinations delivered by the open platform. The chapter presents a risk-based approach for selecting and prioritizing test cases to test service-based systems. The problem addressed is in the context of semantic web services. Such services introduce semantics to service integration and interoperation using ontology models and specifications like OWL-S. They are considered to be the future in World Wide Web evolution. However, due to typically complex ontology relationships, semantic errors are more difficult to detect, compared with syntactic errors. The models described in the chapter analyse semantics from various perspectives such as ontology dependency, ontology usage and service workflow, in order to identify factors that contribute to risks in the delivery of these services. Risks are analysed from two aspects, namely failure probability and importance, and three layers: ontology data, specific services and composite services. With this approach, test cases are associated to the semantic features and schedule test execution on the basis of risks of their target features. Risk assessment is then used to control the process of web services progressive group testing, including test case ranking, test case selection and service ruling out. The chapter presents key techniques used to enable an effective adaptation mechanism: adaptive measurement and adaptation rules. As a statistical testing technique, the approach aims to detect, as early as possible, the problems with highest impact on the users. A number of examples are used to illustrate the approach.

Part III: Operational Risk Analytics

The data described in Part II requires specialized analytics in order to become information and in order for that information to be turned, in a subsequent phase of its analysis, into knowledge. These analytics will be described here.

Chapter 7:   *Scoring models for operational risks* (P. Giudici)

This chapter deals with the problem of analysing and integrating qualitative and quantitative data. In particular it shows how, on the basis of the experience and opinions of internal company 'experts', a scorecard is derived, producing a ranking of different risks and a prioritized list of improvement areas and related controls. Scorecard models represent a first step in risk analysis. The chapter presents advanced approaches and statistical models for implementing such models.

Chapter 8:   *Bayesian merging and calibration for operational risks* (S. Figini)

According to the Basel II Accord, banks are allowed to use the advanced measurement approach (AMA) option for the computation of their capital charge covering operational risks. Among these methods, the loss distribution approach (LDA) is the most sophisticated one. It is highly risk sensitive as long as internal data is used in the calibration process. Given that, LDA is more closely related to the actual risks of each bank. However, it is now widely recognized that calibration on internal data only is not sufficient for computing accurate capital requirements. In other words, internal data should be supplemented with external data. The goal of the chapter is to provide a rigorous statistical method for combining internal and external data and to ensure that merging both databases results in unbiased estimates of the severity distribution.

Chapter 9:   *Measures of association applied to operational risks* (R. Kenett and S. Salini)

Association rules are basic analysis tools for unstructured data such as accident reports, call-centre recordings and customer relationship management (CRM) logs. Such tools are commonly used in basket analysis of shopping carts for identifying patterns in consumer behaviour. The chapter shows how association rules are used to analyse unstructured operational risk data in order to provide risk assessments and diagnostic insights. It presents a new graphical display of association rules that permits effective clustering of associations with a novel interest measure of association rule called the relative linkage disequilibrium.

Part IV: Operational Risk Applications and Integration with other Disciplines

Operational risk management is not a stand-alone management discipline. This part of the book demonstrates how operational risk management relates to other management issues and intelligent regulatory compliance.

Chapter 10:   *Operational risk management beyond AMA: new ways to quantify non-recorded losses* (G. Aprile, A. Pippi and S. Visinoni)

A better understanding of the impact of IT failures on the overall process of operational risk management can be achieved not only by looking at the risk events with a bottom line effect, but also by drilling down to consider the potential risks in terms of missed business opportunities and/or near losses. Indeed, for banking regulatory purposes, only events which are formally accounted for in the books are considered when computing the operational capital at risk. Yet, the 'hidden' impact of operational risks is of paramount importance under the implementation of the Pillar 2 requirements of Basel II, which expands the scope of the analysis to include reputation and business risk topics. This chapter presents a new methodology in operational risk management that addresses these issues. It helps identify multiple losses, opportunity losses and near misses, and quantifies their potential business impact. The main goals are: (1) to reconstruct multiple-effect losses, which is compliant with Basel II requirements; and (2) to quantify their potential impact due to reputation and business risks (opportunity losses) and low-level events (near misses), which is indeed a possible extension to the Basel II advanced measurement approach (AMA). As a consequence, the proposed methodology has an impact both on daily operations of a bank and at the regulatory level, by returning early warnings on degraded system performance and by enriching the analysis of the risk profile beyond Basel II compliance.

Chapter 11: *Combining operational risks in financial risk assessment scores* (M. Munsch, S. Rohe and M. Jungemann-Dorner)

The chapter's central thesis is that efficient financial risk management must be based on an early warning system monitoring risk indicators. Rating and scoring systems are tools of high value for proactive credit risk management and require solid and carefully planned data management. The chapter introduces a business retail rating system based on the Creditreform solvency index which allows a fast evaluation of a firm's creditworthiness. Furthermore, it evaluates the ability of quantitative financial ratings to predict fraud and prevent crimes like money laundering. This practice-oriented approach identifies connections between typical financing processes, operational risks and risk indicators, in order to point out negative developments and trends, enabling those involved to take remedial action in due time and thereby reverse these trends.

Chapter 12: *Intelligent regulatory compliance* (M. Spies, R. Gubser and M. Schacher)

In view of the increasing needs for regulation of international markets, many regulatory frameworks are being defined and enforced. However, the complexity of the regulation rules, frequent changes and differences in national legislation make it extremely complicated to implement, check or even prove regulatory compliance of company operations or processes in a large number of instances. In this context, the Basel II framework for capital adequacy (soon to evolve to Basel III) is currently being used for defining internal assessment processes in banks and other financial services providers. The chapter shows how recent standards and specifications related to business vocabularies and rules enable intelligent regulatory compliance (IRC). IRC is taken to mean semi-automatic or fully automated procedures that can check business operations of relevant complexity for compliance against a set of rules that express a regulatory standard. More specifically, the BMM (Business Motivation Model) and SBVR (Semantics of Business Vocabularies and business Rules) specifications by the Object Management Group (OMG) provide a formal basis for representing regulation systems in a sufficiently formal way to enable IRC of business processes. Besides the availability of automatic reasoning systems, IRC also requires semantics-enabled analysis of business service and business performance data such as process execution logs or trace data. The MUSING project contributed several methods of analysis to the emerging field of IRC. The chapter discusses standards and specifications for business governance and IRC based on BMM and SBVR.

Chapter 13:   *Democratisation of enterprise risk management* (P. Lombardi, S. Piscuoglio, R. Kenett, Y. Raanan and M. Lankinen)

This chapter highlights the interdisciplinary value of the methodologies and solutions developed for semantically enhanced handling of operational risks. The three domains dealt with are operational risk management, financial risk management and internationalization. These areas are usually treated as 'worlds apart' because of the distance of the players involved, from financial institutions to public administrations, to specialized consultancy companies. This proved to be fertile common ground, not only for generating high-value tools and services, but also for a 'democratised' approach to risk management, a technology of great importance to SMEs worldwide.

Chapter 14:   *Operational risks, quality, accidents and incidents* (R. Kenett and Y. Raanan)

This concluding chapter presents challenges and directions for operational risk management. The first section provides an overview of a possible convergence between risk management and quality management. The second section is based on a mapping of uncertainty behaviour and decision-making processes due to Taleb (2007). This classification puts into perspective so-called 'black swans', rare events with significant impact. The third section presents a link between management maturity and the application of quantitative methods in organizations. The fourth section discusses the link between accidents and incidents and the fifth section is a general case study from the oil and gas industry. This illustrates the applicability of operational risk management to a broad range of industries. A final summary section discusses challenges and opportunities in operational risks. Chapter 14 refers throughout to previous chapters in order to provide an integrated view of the material contained in the book.

The book presents state-of-the-art methods and technology and concrete implementation examples. Its main objective is to push forward the operational risk management envelope in order to improve the handling and prevention of risks. It is hoped that this work will contribute, in some way, to organizations which are motivated to improve their operational risk management practices and methods with modern technology. The potential benefits of such improvements are immense.

Notes on Contributors

Ron S. Kenett

Ron Kenett is Chairman and CEO of KPA Ltd (an international management consulting firm with head office in Raanana, Israel), Research Professor at the University of Turin and International Professor Associate at the Center for Risk Engineering of NYU Poly, New York. He has over 25 years of experience in restructuring and improving the competitive position of organizations by integrating statistical methods, process improvements, supporting technologies and modern quality management systems. For 10 years he served as Director of Statistical Methods for Tadiran Telecommunications Corporation and, previously, as researcher at Bell Laboratories in New Jersey. His 160 publications and seven books are on topics in industrial statistics, multivariate statistical methods, improvements in system and software development and quality management. He is Editor in Chief, with F. Ruggeri and F. Faltin, of the *Encyclopedia of Statistics in Quality and Reliability* (John Wiley & Sons, Inc., 2007) and of the international journal *Quality Technology and Quantitative Management*. He was President of ENBIS, the European Network for Business and Industrial Statistics, and his consulting clients include hp, EDS, SanDisk, National Semiconductors, Cisco, Intel, Teva, Merck Serono, Perrigo, banks, healthcare systems, utility companies and government organizations. His PhD is in mathematics from the Weizmann Institute of Science, Rehovot, Israel, and BSc in mathematics, with first-class honours, from Imperial College, London University.

Yossi Raanan

Yossi Raanan is a Senior Consultant and Strategic Partner in KPA, Ltd and Senior Lecturer at the Business School of the College of Management – Academic Studies in Rishon LeZion, Israel. He is also a former dean of that school. He has extensive experience in areas of information technology, data and computer communications and quality management as well as in applying management concepts, tools and know-how to realistic problems, creating applicable, manageable solutions that improve business performance and profitability. His publications and conference talks mirror this knowledge. In addition, he has served on the board of directors of a leading mutual trust company and served as the head of its investment committee, as the chairman of the board of a government-owned company, and a director of a number of educational institutions and a number of start-up technology companies. His PhD is in operations research from Cornell University, Ithaca, NY, with a dissertation on ground-breaking applications of the theories of Lloyd S. Shapley and Robert J. Aumann (Nobel laureate in Economics, 2005) to real-life situations. His BSc, *cum laude*, is in mathematics from the Hebrew University, Jerusalem.

Giorgio Aprile

Giorgio Aprile is head of the operational risk management function in the Monte dei Paschi di Siena Banking Group, the third banking group in Italy and the second certified for an AMA model. He started his activities in OpR in 2003, and from 2005 he has been responsible for implementing the Advanced Measuring Approach (AMA) in the MPS group; the AMA model has been fully running since January 2007. He was born in 1971 and graduated in nuclear engineering in 1996 with a dissertation thesis on the safety analysis of Ukrainian nuclear power plants. He worked as a risk analyst in the oil and gas industry for five years, on different research projects, mainly focused on the 'smart' use of field data to reduce operational risks in offshore plants and the prevention of environmental disasters. At the end of 2001 he joined the MPS group and focused on the development of advanced tools for risk management and strategic marketing for the bank. From June 2005 he has been the head of the OpR management function.

Xiaoying Bai

Xiaoying Bai is currently an Associate Professor at the Department of Computer Science and Technology of Tsinghua University. She received her PhD degree in computer science in 2001 from Arizona State University in the United States. After that, she joined the Department of Computer Science and Technology of Tsinghua University in 2002 as an Assistant Professor and was promoted to Associated Professor in 2005. Her major research area is software engineering, especially model-driven testing and test automation techniques in various software paradigms such as distributed computing, service-oriented architecture and embedded systems. She has led more than 10 projects funded by the National Key Science and Technology Project, National Science Foundation and National High Tech 863 Program in China, as well as international collaboration with IBM and Freescale. She was also involved as a key member of the Key Project of Chinese National Programs for Fundamental Research and Development (973 Program). She has published over 50 papers in journals and international conference proceedings of ANSS, COMPSAC, ISADS, SIGSOFT CBSE, ICWS, etc. She is the co-author of a Chinese book, *Service-Oriented Software Engineering*. She was the Program Chair of the first IEEE International Conference on Service Oriented System Engineering and is now serving or has served as PC member for many software engineering conferences, including ICEBE, QSIC, COMPSAC, SEKE, HASE, WISE, etc., and as a reviewer for international journals.

Kalina Bontcheva

Kalina Bontcheva is a Senior Researcher at the Natural Language Processing Laboratory of the University of Sheffield. She obtained her PhD from the University of Sheffield in 2001 and has been a leading developer of GATE since 1999. She is Principal Investigator for the TAO and MUSING projects where she coordinates works on ontology-based information extraction and ontology learning.

Thierry Declerck

Thierry Declerck (MA in philosophy, Brussels; MA in computer linguistics, Tübingen) is a Senior Consultant at the DFKI Language Technology Lab. Before joining DFKI in 1996, he worked at the Institute of Natural Language Processing (IMS) in Stuttgart. At DFKI, he worked first in the field of information extraction. He was later responsible for the EU FP5 project MUMIS (MUltiMedia Indexing and Searching). He has also worked at the University of Saarland, conducting two projects, one on a linguistic infrastructure for e-Content and the other one, Esperonto, on the relation between NLP and the Semantic Web. He has actually led the DFKI contribution to the European Network of Excellence 'K-Space' (Knowledge Space of semantic inference for automatic annotation and retrieval of multimedia content, see www.k-space.eu) and is coordinating the research work packages of the IP MUSING (see www.musing.eu), both projects being part of the 6th Framework Programme in IST. He is also actively involved in standardization activities in the context of ISO TC37/SC4.

Silvia Figini

Silvia Figini has a PhD in statistics from Bocconi University in Milan. She is a researcher in the Department of Statistics and Applied Economics L. Lenti, University of Pavia, a member of the Italian Statistical Society and author of publications in the area of methodological statistics, Bayesian statistics and statistical models for financial risk management. She teaches undergraduate courses in applied statistics and data mining.

Paolo Giudici

Paolo Giudici is Professor at the University of Pavia where he is a lecturer in data analysis, business statistics and data mining, as well as risk management (at Borromeo College). He is also Director of the Data Mining Laboratory; a member of the University Assessment Board; and a coordinator of the Institute of Advanced Studies school on 'Methods for the management of complex systems'. He is the author of 77 publications, among which are two research books and 32 papers in Science Citation Index journals. He has spent several research periods abroad, in particular at the University of Bristol, the University of Cambridge and at the Fields Institute (Toronto) for research in the mathematical sciences. He is the coordinator of two national research grants: one (PRIN, 2005–2006) on 'Data mining methods for e-business applications'; and one (FIRB 2006–2009) on 'Data mining methods for small and medium enterprises'. He is also the local coordinator of a European integrated project on 'Data mining models for advanced business intelligence applications' (MUSING, 2006–2010) and responsible for the Risk Management Interest Group of the European Network for Business and Industrial Statistics. He is also a member of the Italian Statistical Society, the Italian Association for Financial Risk Management and the Royal Statistical Society.

Valerio Grossi

Valerio Grossi holds a PhD in computer science from the University of Pisa. Currently, he is a Research Associate at the Department of Computer Science, University of Pisa, in the business intelligence research area. He was involved in the MUSING project on the development of strategies of operational risk management. His research activities include machine learning, data mining and knowledge discovery with special reference to mining data streams.

Rolf Gubser

Rolf Gubser graduated in computer science in 1990. He is a founding member of KnowGravity, Inc., a leading contributor to several OMG specifications like BMM (Business Motivation Model), SBVR (Semantics of Business Vocabulary and business Rules) and BPMN (Business Process Modelling Notation). Based on those specifications, he focuses on developing and applying Model Driven Enterprise Engineering™, a holistic approach that integrates strategic planning, risk and compliance management, business engineering, as well as IT support. Before KnowGravity, he worked as a Senior Consultant at NCR and Born Informatik AG.

Monika Jungemann-Dorner

Monika Jungemann-Dorner, born in 1964, has a Masters degree in linguistics and economics. She has been working on international projects for more than 10 years. At the German Chamber Organization she was responsible for the management and deployment of a number of innovative European projects (e.g. knowledge management and marketplaces). Since January 2004, she has worked as Senior International Project Manager for the Verband der Vereine Creditreform eV, Germany.

Hans-Ulrich Krieger

Hans-Ulrich Krieger is a Senior Researcher at the German Research Centre for Artificial Intelligence (DFKI). He studied computer science and physics at the RWTH Aachen and received a PhD (Dr. rer. nat.) in computer science from Sarland University in 1995. His global research in computational linguistics has focused on linguistic formalisms, their efficient implementation and their mathematical foundations. He has worked on the morpho-syntax interface in HPSG, was the prime contributor of the typed description language TDL and the typed shallow processor SProUT, and led the group on deep linguistic processing in the Verbmobil project. His latest work in computational linguistics concerns the compilation of constraint-based grammar formalisms into weaker frameworks and the integration of deep and shallow processing methods. During the last few years, he has worked at the intersection of linguistic processing, general (world) knowledge representation and inference systems in the AIRFORCE, COLLATE and MUSING projects. He has implemented a time ontology that addresses temporal granularity and temporal underspecification in natural language and devised a general scheme for incorporating time into arbitrary description logic ontologies. His expertise in Semantic Web technologies includes theoretical frameworks, such as OWL, SWRL and SPARQL, as well as practical systems, like RACER, Pellet, OWLIM and Jena.

Markus Lankinen

Markus Lankinen, MSc (Economics), Vaasa, 1996, is Managing Director for the St-Petersburg-based consultancy company ManNet Partners and Regional Manager of the Finnish company FCG Planeko Ltd. As a management consultant, he is active in international business planning, specializing in start-up and development strategies. He is a member of the Finnish Association of Business School Graduates.

Christian Leibold

Christian Leibold holds a Diploma (master equivalent) of Computer Science from Munich Ludwig-Maximilians-University. The topic of his diploma thesis was the evaluation of content repository solutions and implementation at BMW Group Design Project Management. During his studies he concentrated on system architectures, specializing in Java EE and user interaction, including the teaching of tutorials accompanying the lectures in spoken dialogue systems and enterprise integration. He joined STI Innsbruck in April 2007. Since then he has won a research grant for the market preparation of a research prototype on the classification of information. In the EC IP MUSING he has a major role as project researcher. Before joining STI he was involved in industrial projects with BMW Group, Infineon Technologies, Take2 Interactive Software and Execcon Consulting, covering various parts of the engineering loop of software and hardware products (e.g. requirements analysis, implementation, quality assurance, project management and process development).

Paolo Lombardi

Paolo Lombardi is Head of Internet Banking at Monte dei Paschi di Siena, the third largest Italian banking group. He joined the Montepaschi group in 2001, managing initiatives on financial-services-related innovation. From 2004, within the Organization Department, he undertook programme management responsibilities in the framework of the group's Basel II initiative. Since April 2007 he has also been responsible for the group's Contact Centre. He holds a *summa cum laude* Masters degree in nuclear engineering from the University of Pisa (1989), and has been appointed as Post-Graduate Researcher at the University of California Santa Barbara. He has been involved in the design and execution of European Commission research and technological development projects since 1992, and collaborates with the EC's DG Information Society as an evaluator and a reviewer.

Michael Munsch

Michael Munsch studied finance at the Comprehensive University of Essen. He received his PhD in the field of international finance risk management. Since August 2000 he has been Executive Manager of Creditreform Rating AG. He was previously Head of the Department of Risk Management at the Verband der Vereine Creditreform eV in Neuss, where he coordinated the development of new systems for credit rating and risk management. After finishing his studies in business administration he became a consultant for business clients at an international bank and was employed in the Department of Finance of a large international company. From 1989 to 1995 he worked in the Department of Finance and Financial Accounting at the University of Essen.

Richard Pike

Richard Pike, CCH SWORD Product Director, has more than 15 years' experience in risk management and treasury IT. He has analysed, designed and managed the development of core risk management systems for large international financial institutions. He was recently chosen as one of the 50 most influential people in operational risk, by *Operational Risk & Compliance* magazine. He is a regular speaker and writer on risk management issues.

Antonio Pippi

Antonio Pippi is a telecommunications engineer. In 2001–2005 he was with the Department of Information Engineering at the University of Siena, where he obtained a PhD with a dissertation on mathematical models for the study of electromagnetic waves. In 2006 he joined Monte dei Paschi di Siena Banking Group, first as an IT project manager in the credit and financial sectors (adoption of Basel II recommendations). Then, in 2007, he moved to the operational risk management function, where he worked on qualitative and quantitative statistical models of operational risks, according to the advanced measurement approach (AMA) of the Basel II framework. In 2009 he joined the Zurich Financial Services Group insurance undertaking, where is currently in the Risk & Control Office of the Italian branch.

Salvatore Piscuoglio

Salvatore Piscuoglio is a member of the Internet Banking Department at Monte dei Paschi di Siena, the third largest Italian banking group. He joined the Montepaschi group in 2003. He was part of the Organization Department until 2009, managing in particular the re-engineering of finance and risk management processes. From 2004 to 2007 he was head of the organizational team both for the development of the advanced measuring approach (AMA) for operational risk and for the implementation of the internal model for market risk. He holds a BA degree, with honours, in economics from the 'Federico II' University in Naples (2000) and a Masters degree in economics and banking from the University of Siena (2002).

Silvia Rohe

Silvia Rohe completed her professional training at a technical college for bank management and business economy. She also studied banking services and operations at Banker's Academy (Bankakademie) in Cologne. She began her professional career as a credit analyst with a major Swiss bank. From 1989 to 2006 she was employed by a prominent automotive bank as Group Manager of the Risk Management Department. Since January 2007, she has been a Senior Consultant for Verband der Vereine Creditreform eV, in Neuss.

Andrea Romei

Andrea Romei holds a PhD in computer science from the University of Pisa (2009). Currently, he is a Research Associate and a member of the KDDLab at the Department of Computer Science (University of Pisa). His research interests include data mining, inductive databases, business intelligence, web mining, knowledge representation and knowledge discovery, parallel and distributed data mining. His main contribution has been on the design and implementation of an inductive database system from XML data.

Horacio Saggion

Horacio Saggion is a Research Fellow in the Natural Language Processing Group, Department of Computer Science, University of Sheffield. He obtained his PhD in

2000 from Université de Montréal, his Masters degree (1995) from Universidade Estadual de Campinas (UNICAMP), and his undergraduate degree of 'Licenciado' in 1988 from Universidad de Buenos Aires. He has published over 50 works in conference, workshop and journal papers as well as writing three book chapters. Together with his research career, he has been an active teacher at Universidad de Buenos Aires, Universidad Nacional de Quilmes and Université de Montréal. He has been a member of several scientific programme committees in natural language processing and artificial intelligence.

Silvia Salini

Silvia Salini holds a degree in statistics from Catholic University of Milan and PhD in statistics from University of Milan Bicocca. Currently she is Assistant Professor of Statistics at the Department of Economics, Business and Statistics of the University of Milan. Her main research interests are multivariate statistical analysis, data mining and statistical methods in social science.

Markus Schacher

Markus Schacher is Co-founder and a Senior Consultant of KnowGravity Inc., Switzerland. He has been teaching business-oriented specification techniques for more than 20 years and offered the first UML courses in Switzerland back in early 1997. As a specialist in knowledge-based systems and technologies, he provides training and consulting on the Business Rules Approach, executable UML (xUML) and OMG's Model Driven Architecture (MDA). He is an active member of the Business Rules Group (www.businessrulesgroup.org) and a member of the Business Rules Team that developed the OMG specifications 'Semantics of Business Vocabulary and Business Rules (SBVR)' and 'Business Motivation Model (BMM)'. He co-authored the first comprehensive book on the Business Rules Approach written in German, which served as the basis for the development of the Model Driven Enterprise Engineering™ framework that integrates the major business-oriented OMG specifications with OMG's IT-oriented specifications. Finally, he is the lead architect of KnowGravity's KnowEnterprise(r) and CASSANDRA platforms that implement a number of these OMG specifications.

Marcus Spies

Marcus Spies, born in 1956, studies cognitive science, computer science and economics at Berlin Technical University and University of Hagen. His PhD thesis was on uncertainty management in knowledge-based systems 1988. From 1987 to 1989 he was a Postdoctoral Fellow at IBM Science Centre Heidelberg. From 1989 to 1995, he was a research staff member at IBM, undertaking applied research in statistical expert systems for environmental monitoring and analysis, with a successful patent application in language modelling for IBM's speech recognition system. From 1996 to 2001, he was a Senior Consultant at IBM, carrying out consulting engagements with key IBM customers in software project management, IT security management and knowledge management. Since 2001 he has been Professor of Knowledge Management at LMU, University of Munich, and from 2006 to 2008, Visiting Professor at Digital Enterprise Research Institute, University of Innsbruck. In 2007 he was Programme Chair, and in 2008 General Chair, of the IEEE Enterprise Distributed Object Conferences in Annapolis and Munich. Since 2008 he has been Scientific and Technical Director of the EU integrated MUSING project (www.musing.eu). He has written more than 55 publications, among them a monograph on uncertainty management in expert systems and an introductory textbook on propositional calculus and first-order logic with applications to knowledge representation (both books in German).

Stefano Visinoni

Stefano Visinoni is an electronic engineer. In 2000–2007 he was Senior Consultant at Accenture, Business Process Outsourcing Unit, in the IT-Finance Team where he mainly dealt with IT risk management. In 2007 he joined Antonveneta ABN-AMRO Banking Group, as Risk Manager, and then, in 2008, he joined Monte dei Paschi di Siena (MPS) Banking Group in the operational risk management function. There, he mainly deals with qualitative models for the measurement and management of operational risks. Since 2007 he has held a position with the Department of Economic Science at the University of Verona, where he teaches the Business Process Outsourcing course in the Bank and Insurance IT Environments degree programme.

List of Acronyms

AAWS	Amazon Associate Web Services
AI	Artificial Intelligence
AMA	Advanced Measurement Approach
API	Application Programming Interface
AQI	Air Quality Index
BACH	Bank for the Accounts of Companies Harmonized
BCBS	Basel Committee on Banking Supervision
BEICF	Business Environment and Internal Control Factor
BI	Business Intelligence
BIA	Basic Indicator Approach
BIS	Bank of International Settlements
BL	Business Line
BMM	Business Motivation Model
BN	Bayesian Network
BPEL	Business Process Execution Language
BPMN	Business Process Modelling Notation
BSC	Balanced ScoreCard
CEO	Chef Executive Officer
cGMP	current Good Manufacturing Practices
COSO	Committee of Sponsoring Organizations of the Treadway Commission
CRF	Conditional Random Field
CRM	Customer Relationship Management
CRO	Chief Risk Officer
DB	DataBase
DBMS	DataBase Management System
DFU	Definerte fare- og ulykkessituasjoner (Defined Hazard For Accidents)
DG	Dependence Graph
DoE	Design of Experiments
DRP	Disaster Recovery Plan
EC	Event Category
ERM	Enterprise(-wide) Risk Management
ERP	Enterprise Resource Planning

ET	Event Type
ETL	Extract, Transform and Load
FDA	Food and Drug Administration
FRM	Financial Risk Management
FSI	Financial Services and Insurance
GATE	General Architecture for Text Engineering
GUI	Graphical User Interface
HLT	Human Language Technology
HMM	Hidden Markov Model
HTML	Hyper Text Markup Language
ICT	Information and Communications Technology
IE	Information Extraction
InfoQ	Information Quality
IRC	Intelligent Regulatory Compliance
IRCSE	Integrated Risk Check-up System for Enterprises
IT	Information Technology
JAPE	Java Annotation Pattern Engine
KPI	Key Performance Indicator
KRI	Key Risk Indicator
KYC	Know Your Customer
LDA	Loss Distribution Approach
LDCE	Loss Data Collection Exercise
LHS	Left hand Side
LR	Language Resource
MDA	Model-Driven Architecture
MUC	Message Understanding Conference
MUSING	MUlti-industry, Semantic-based next generation business INtelliGence
NACE	Nomenclature statistique des Activités économiques dans la Communauté Européenne
NLP	Natural Language Processing
OAT	Ontology Annotation Tool
OBIE	Ontology-Based Information Extraction
OLAP	OnLine Analytical Processing
OMG	Object Management Group
OpR	Operational Risk management
OU	Organizational Unit
OWL	Web Ontology Language
PBX	Private Branch eXchange
PD	Probability of Default
PDF	Portable Document Format
POS	Part of Speech
PR	Processing Resource
RCA	Risk and Control Assessment

RDF	Resource Description Framework
RDF(S)	RDF Schema
RHS	Right Hand Side
RLD	Relative Linkage Disequilibrium
RTF	Rich Text Format
SBVR	Semantics of Business Vocabularies and business Rules
SGML	Standard Generalized Markup Language
SLA	Service Level Agreement
SME	Small and Medium-sized Enterprise
SOA	Service-Oriented Architecture
SQL	Standard Query Language
SVM	Support Vector Machine
UML	Unified Modeling Language
URI	Uniform Resource Identifier
URL	Uniform Resource Locator
VaR	Value at Risk
VNO	Virtual Network Operator
VR	Visualization resource
W3C	World Wide Web Consortium
WS	Web Services
WSDL	Web Service Description Language
XBRL	eXtensible Business Reporting Language
XML	eXtensible Markup Language

Part I

INTRODUCTION TO OPERATIONAL RISK MANAGEMENT

1

Risk management: a general view

Ron S. Kenett, Richard Pike and Yossi Raanan

1.1 Introduction

Risk has always been with us. It has been considered and managed since the earliest civilizations began. The Old Testament describes how, on the sixth day of creation, the Creator completed his work and performed an *ex post* risk assessment to determine if further action was needed. At that point in time, no risks were anticipated since the 31st verse of Genesis reads 'And God saw every thing that he had made, and, behold, it was very good' (Genesis 1: 31).

Such evaluations are widely conducted these days to determine risk levels inherent in products and processes, in all industries and services. These assessments use terms such as 'probability or threat of a damage', 'exposure to a loss or failure', 'the possibility of incurring loss or misfortune'. In essence, risk is linked to uncertain events and their outcomes. Almost a century ago, Frank H. Knight proposed the following definition:

> Risk is present where future events occur with measureable probability.

Quoting more from Knight:

> Uncertainty must be taken in a sense radically distinct from the familiar notion of risk, from which it has never been properly separated . . . .

Operational Risk Management: A Practical Approach to Intelligent Data Analysis Edited by Ron S. Kenett and Yossi Raanan © 2011 John Wiley & Sons, Ltd

The essential fact is that 'risk' means in some cases a quantity susceptible of measurement, while at other times it is something distinctly not of this character; and there are far-reaching and crucial differences in the bearings of the phenomena depending on which of the two is really present and operating.... It will appear that a measurable uncertainty, or 'risk' proper, as we shall use the term, is so far different from an unmeasurable one, that it is not in effect an uncertainty at all'.

(Knight, 1921)

According to Knight, the distinction between risk and uncertainty is thus a matter of knowledge. Risk describes situations in which probabilities are available, while uncertainty refers to situations in which the information is too imprecise to be summarized by probabilities. Knight also suggested that uncertainty can be grasped by an 'infinite intelligence' and that to analyse these situations theoreticians need a continuous increase in knowledge. From this perspective, uncertainty is viewed as a lack of knowledge about reality.

This separates 'risk' from 'uncertainty' where the probability of future events is not measured. Of course what are current uncertainties (e.g. long-range weather forecasts) may some day become risks as science and technology make progress.

The notion of risk management is also not new. In 1900, a hurricane and flood killed more than 5000 people in Texas and destroyed the city of Galveston in less than 12 hours, materially changing the nature and scope of weather prediction in North America and the world. On 19 October 1987, a shock wave hit the US stock market, reminding all investors of the inherent risk and volatility in the market. In 1993, the title of 'Chief Risk Officer' was first used by James Lam, at GE Capital, to describe a function to manage 'all aspects of risk' including risk management, back-office operations, and business and financial planning. In 2001, the terrorism of September 11 and the collapse of Enron reminded the world that nothing is too big to collapse.

To this list, one can add events related to 15 September 2008, when Lehman Brothers announced that it was filing for Chapter 11 bankruptcy protection. Within days, Merrill Lynch announced that it was being sold to rival Bank of America at a severely discounted price to avert its own bankruptcy. Insurance giant AIG, which had previously received an AAA bond rating (one of only six US companies to hold an AAA rating from both Moody's and S&P) stood on the brink of collapse. Only an $85 billion government bailout saved the company from experiencing the same fate as Lehman Brothers. Mortgage backers Fannie Mae and Freddie Mac had previously been put under federal 'governorship', to prevent the failure of two major pillars in the US mortgage system. Following these events, close to 1000 financial institutions have shut down, with losses up to $3600 billion.

The car industry has also experienced such events. After Toyota announced a recall of 2.3 million US vehicles on 21 January 2010, its shares dropped 21%, wiping out $33 billion of the company's market capitalization. These widely publicized events keep reinvigorating risk management.

The Food and Drug Administration, National Aeronautics and Space Administration, Department of Defense, Environmental Protection Agency, Securities and Exchange Commission and Nuclear Regulatory Commission, among others, have all being implementing risk management for over a decade. Some basic references that form the basis for these initiatives include: Haimes (2009), Tapiero (2004), Chorafas (2004), Ayyub (2003), Davies (1996) and Finkel and Golding (1994).

Risk management, then, has long been a topic worth pursuing, and indeed several industries are based on its successful applications, insurance companies and banks being the most notable. What gives this discipline enhanced attention and renewed prominence is the belief that nowadays we can do a better job of it. This perception is based on phenomenal developments in the area of data processing and data analysis. The challenge is to turn 'data' into information, knowledge and deep understanding (Kenett, 2008). This book is about meeting this challenge. Many of the chapters in the book are based on work conducted in the MUSING research project. MUSING stands for MUlti-industry, Semantic-based next generation business INtelliGence (MUSING, 2006). This book is an extended outgrowth of this project whose objectives were to deliver next generation knowledge management solutions and risk management services by integrating Semantic Web and human language technologies and to combine declarative rule-based methods and statistical approaches for enhancing knowledge acquisition and reasoning. By applying innovative technological solutions in research and development activities conducted from 2006 through 2010, MUSING focused on three application areas:

1. *Financial risk management.* Development and validation of next generation (Basel II and beyond) semantic-based business intelligence (BI) solutions, with particular reference to credit risk management and access to credit for enterprises, especially small and medium-sized enterprises (SMEs).

2. *Internationalization.* Development and validation of next generation semantic-based internationalization platforms supporting SME internationalization in the context of global competition by identifying, capturing, representing and localizing trusted knowledge.

3. *Operational risk management.* Semantic-driven knowledge systems for operational risk measurement and mitigation, in particular for IT-intensive organizations. Management of operational risks of large enterprises and SMEs impacting positively on the related user communities in terms of service levels and costs.

Kenett and Shmueli (2009) provide a detailed exposition of how data quality, analysis quality and information quality are all required for achieving knowledge with added value to decision makers. They introduce the term InfoQ to assess the quality of information derived from data and its analysis and propose several practical ways to assess it. The eight InfoQ dimensions are:

1. *Data granularity.* Two aspects of data granularity are measurement scale and data aggregation. The measurement scale of the data must be adequate for the purpose of the study and. The level of aggregation of the data should match the task at hand. For example, consider data on daily purchases of over-the-counter medications at a large pharmacy. If the goal of the analysis is to forecast future inventory levels of different medications, when restocking is done on a weekly basis, then we would prefer weekly aggregate data to daily aggregate data.

2. *Data structure.* Data can combine structured quantitative data with unstructured, semantic-based data. For example, in assessing the reputation of an organization one might combine data derived from balance sheets with data mined from text such as newspaper archives or press reports.

3. *Data integration.* Knowledge is often spread out across multiple data sources. Hence, identifying the different relevant sources, collecting the relevant data and integrating the data directly affects information quality.

4. *Temporal relevance.* A data set contains information collected during a certain period of time. The degree of relevance of the data to the current goal at hand must be assessed. For instance, in order to learn about current online shopping behaviours, a data set that records online purchase behaviour (such as Comscore data, www.comscore.com) can be irrelevant if it is even one year old, because of the fast-changing online shopping environment.

5. *Sampling bias.* A clear definition of the population of interest and how a sample relates to that population is necessary in both primary and secondary analyses. Dealing with sampling bias can be proactive or reactive. In studies where there is control over the data acquisition design (e.g. surveys), sampling schemes are selected to reduce bias. Such methods do not apply to retrospective studies. However, retroactive measures such as post-stratification weighting, which are often used in survey analysis, can be useful in secondary studies as well.

6. *Chronology of data and goal.* Take, for example, a data set containing daily weather information for a particular city for a certain period as well as information on the air quality index (AQI) on those days. For the United States such data is publicly available from the National Oceanic and Atmospheric Administration website (www.noaa.gov). To assess the quality of the information contained in this data set, we must consider the purpose of the analysis. Although AQI is widely used (for instance, for issuing a 'code red' day), how it is computed is not easy to figure out. One analysis goal might therefore be to find out how AQI is computed

from weather data (by reverse engineering). For such a purpose, this data is likely to contain high-quality information. In contrast, if the goal is to predict future AQI levels, then the data on past temperatures contains low-quality information.

7. *Concept operationalization.* Observable data is an operationalization of underlying concepts. 'Anger' can be measured via a questionnaire or by measuring blood pressure; 'economic prosperity' can be measured via income or by unemployment rate; and 'length' can be measured in centimetres or in inches. The role of concept operationalization is different for explanatory, predictive and descriptive goals.

8. *Communication and data visualization.* If crucial information does not reach the right person at the right time, then the quality of information becomes poor. Data visualization is also directly related to the quality of information. Poor visualization can lead to degradation of the information contained in the data.

Effective risk management necessarily requires high InfoQ. For more on information quality see Guess (2000), Redman (2007) and Kenett (2008).

We are seeking knowledge and require data in order to start the chain of reasoning. The potential of data-driven knowledge generation is endless when we consider both the increase in computational power and the decrease in computing costs. When combined with essentially inexhaustible and fast electronic storage capacity, it seems that our ability to solve the intricate problems of risk management has stepped up several orders of magnitude higher.

As a result, the position of chief risk officer (CRO) in organizations is gaining popularity in today's business world. Particularly after the 2008 collapse of the financial markets, the idea that risk must be better managed than it had been in the past is now widely accepted (see Kenett, 2009). Still, this position is not easy to handle properly. In a sense it is a new version of the corporate quality manager position which was popular in the 1980s and 1990s. One of the problems inherent in risk management is its almost complete lack of glamour. Risk management done well is treated by most people like electric power or running water – they expect those resources to be ever present, available when needed, inexpensive and requiring very little management attention. It is only when they are suddenly unavailable that we notice them. Risks that were well managed did not materialize, and their managers got little attention. In general, risk management positions provide no avenues to corporate glory. Indeed, many managers distinguish themselves in times of crisis and would have gone almost completely unnoticed in its absence. Fire fighting is still a very prevalent management style. Kenett *et al.* (2008) formulated the Statistical Efficiency Conjecture that stipulates that organizations exercising fire fighting, as opposed to process improvement of quality by design, are less effective in their improvement initiatives. This was substantiated with 21 case studies which were collected and analysed to try to convince management that prevention is carrying significant rewards.

An example of this phenomenon is the sudden glory bestowed on Rudy Giuliani, the former Mayor of New York City, because of his exceptional crisis management in the aftermath of the September 11 terrorist attack on the twin towers. It was enough to launch his bid for the presidency (although not enough, apparently, to get him elected to that office or even to the post of Republican candidate). Had the attacks been avoided, by a good defence intelligence organization, he would have remained just the Mayor of New York City. The people who would have been responsible for the prevention would have got no glory at all, and we might even never have heard about them or about that potential terrible threat that had been thwarted. After all, they were just doing their job, so what is there to brag about? Another reason for not knowing about the thwarted threat, valid also for business risk mitigation strategies, is not exposing the methods, systems and techniques that enabled the thwarting.

Nonetheless, risk management is a critically important job for organizations, much like vaccination programmes. It must be funded properly and given enough resources, opportunities and management attention to achieve concrete results, since it can be critical to the organization's survival. One should not embrace this discipline only after disaster strikes. Organizations should endeavour to prevent the next one by taking calculated, evidence-based, measured steps to avoid the consequences of risk, and that means engaging in active risk management.

1.2   Definitions of risk

As a direct result of risk being a statistical distribution rather than a discrete point, there are two main concepts in risk measurement that must be understood in order to carry out effective risk management:

1. *Risk impact*. The impact (financial, reputational, regulatory, etc.) that will happen should the risk event occur.

2. *Risk likelihood*. The probability of the risk event occurring.

This likelihood usually has a time period associated with it. The likelihood of an event occurring during the coming week is quite different from the likelihood of the same event occurring during the coming year. The same holds true, to some extent, for the risk impact since the same risk event occurring in two different points in time may result in different impacts. These differences between the various levels of impact may even owe their existence to the fact that the organization, realizing that the event might happen, has engaged actively in risk management and, at the later of the two time periods, was better prepared for the event and, although it could not stop it from happening, it succeeded in reducing its impact.

Other base concepts in the risk arena include:

- *Risk event*. An actual instance of a risk that happened in the past.

- *Risk cause*. The preceding activity that triggers a risk event (e.g. fire was caused by faulty electrical equipment sparking).

Risk itself has risk, as measures of risk often are subject to possible change and so measures of risk will often come with a confidence level that tells the reader what the risk of the risk measure is. That is, there may be some uncertainty about the prediction of risk but of course this should never be a reason to avoid the sound practice of risk management, since its application has generated considerable benefits even with less than certain predictions.

1.3   Impact of risk

In her book *Oracles, Curses & Risk Among the Ancient Greeks*, Esther Eidinow shows how the Greeks managed risk by consulting oracles and placing curses on people that affected their lives (Eidinow, 2007). She also posits that risk management is not just a way of handling objective external dangers but is socially constructed and therefore, information about how a civilization perceives risk, provides insights into its social dynamics and view of the world. The type of risks we are concerned with, at a given point in time, also provides insights into our mindset. Specifically, the current preponderance on security, ecological and IT risks would make excellent research material for an anthropologist in 200 years.

This natural tendency to focus on specific types of risk at certain times causes risk issues, as it is exactly the risks you have not been focusing on that can jump up and bite you. In his book *The Black Swan*, Nassim Nicholas Taleb describes events that have a very low probability of occurrence but can have a very great impact (Taleb, 2007). Part of the reasons he gives for these unexpected events is that we have not been focusing on them or their possibilities because of the underlying assumptions we made about our environment (i.e. all swans are white).

It is also true that the impact of many risk events is difficult to estimate precisely, since often one risk event triggers another, sometimes even a chain reaction, and then the measurements tend to become difficult. This distribution of the total impact of a compound event among its components is not of great importance during an initial analysis of risks. We would be interested in the whole, and not in the parts, since our purpose is to prevent the impact. Subsequent, finer, analysis may indeed assign the impacts to the component parts if their happening separately is deemed possible, or if it is possible (and desirable) to manage them separately. A large literature exists on various aspects or risk assessment and risk management. See for example Alexander (1998), Chorafas (2004), Doherty (2000), Dowd (1998), Embrecht *et al.* (1997), Engelmann and Rauhmeier (2006), Jorion (1997), Kenett and Raphaeli (2008), Kenett and Salini (2008), Kenett and Tapiero (2009), Panjer (2006), Tapiero (2004) and Van den Brink (2002).

1.4   Types of risk

In order to mitigate risks the commercial world is developing holistic risk management programmes and approaches under the banner of enterprise risk management (ERM). This framework aims to ensure that all types of risk are considered and attempts are made to compare different risk types within one overall risk measurement approach. There are many ERM frameworks available, but one of the most prevalent is the COSO ERM model created by the Committee of Sponsoring Organizations of the Treadway Commission. This framework categorizes risks within the following types: (1) financial, (2) operational, (3) legal/compliance and (4) strategic.

It is within this framework that this book approaches operational risks. This category is very broad and is present in, and relevant to, all industries and geographies. It covers such diverse topics as IT security, medical malpractice and aircraft maintenance. This diversity means that there are many approaches to measuring operational risk and all differ in terms of quantitative maturity and conceptual rigour. One important scope of the 'operational' category of risks deals with risks that are associated with the operations of information and communications technology (ICT). The reasons for this are that ICT is nowadays a critical component in all enterprises, forming a layer of the business infrastructure, that attracts over half the capital investments of business and thus deserves to be well managed. Moreover, ICT produces diagnostic data that makes tracking, analysing and understanding risk events easier. This encourages getting insights into the causes of risk events and improving their management. These aspects of risk were the focus of the MUSING European Sixth Framework Programme (MUSING, 2006).

1.5 Enterprise risk management

ERM is a holistic approach that views all the areas of risk as parts of an entity called risk. In addition to the fact that the division of risks across the various categories listed above requires tailored decisions, what one organization may call strategic, may be considered operational in another. The view is that the classification into such areas is an important tool to help decompose a very large problem into smaller pieces. However, all these pieces must be dealt with and then looked at by a senior manager in order to determine which risks are dealt with first, which later and which will currently be knowingly ignored or perhaps accepted without any action to manage them.

The basic creed of ERM is simple: 'A risk, once identified, is no longer a risk – it is a management problem.' Indeed, a telling phrase, putting the responsibility and the accountability for risk management and its consequences right where they belong – on the organization's management. It is based on the realization that the issue of what type a risk is – while relevant to the handling of that risk – is totally immaterial when it comes to damages resulting from that risk. Different types of risks may result in similar damages to the organization.

Therefore, the decomposition of risks into separate areas by their functional root causes is no more than a convenience and not an inherent feature of risk. As a result, all risk management efforts, regardless of their functional, organizational or geographical attributes, should be handled together. They should not be treated differently just because of expediency or because some functional areas have 'discovered' risk – sometime disguised by other terms – sooner than other areas. For example, just because accounting deals with financial exposure does not mean that risk management should be subjugated to that functional area. For example the fact that IT departments have been dealing with disaster recovery planning (DRP) to their own installations and services does not mean that risk management belongs in those departments. Risk management should be a distinct activity of the organization, located organizationally where management and the board of directors deem best, and this activity should utilize the separate and important skills deployed in each department – be it accounting, IT or any other department – as needed.

1.6   State of the art in enterprise risk management

A well-established concept that has been deployed across different industries and situations is the concept of three lines of defence. It consists of:

- *The business*. The day-to-day running of the operation and the front office.
- *Risk and compliance*. The continual monitoring of the business.
- *Audit*. The periodic checking of risk and compliance.

This approach has offered thousands of organizations a solid foundation upon which to protect themselves against a range of potential risks, both internal and external. Some organizations adopted it proactively on their own, as part of managing risk, and others may have had it forced upon them through regulators' insistence on external audits.

Regardless of circumstance, the three lines of defence concept is reliable and well proven, but it needs to be periodically updated. Otherwise, its ability to meet the rigours of today's market, where there is an increasing number of risks and regulations, and an ever-increasing level of complexity, becomes outdated.

For the three lines of defence to succeed, the communication and relationship between them needs to be well defined and coordination across all three lines must be clearly established. This is not easy to accomplish. In the majority of organizations, management of the various forms of risk – operational risk, compliance risk, legal risk, IT risk, etc. – is carried out by different teams, creating a pattern of risk silos. Each form of risk, or risk silo, is managed in a different way. This situation leads to a number of negative consequences described below.

1.6.1   The negative impact of risk silos

1.6.1.1   Inefficiency multiplies across silos

Silos may be very efficient at one thing, but that may be at the expense of the overall organization's efficiency. In the case of risk silos, each gathers the information it needs by asking the business managers to provide various information relating to their daily operations and any potential risks associated with them. Because of the silo structure, the business will find itself being asked for this same information on multiple occasions by a multiple of risk silos. These duplicative efforts are inefficient and counterproductive, and lead to frustrated front-office staff disinclined to engage with risk management in the future. The level of frustration is such today that when the recently appointed CEO of a large company asked his senior managers what single change would make their life easier, the reply was to do something to stop the endless questionnaires and check sheets that managers were required to fill out to satisfy risk managers and compliance officers. Frustration among business managers is never a positive development. But it can fully undermine a company's risk management programme as buy-in from the staff is essential.

1.6.1.2   Inconsistency adds to risks

Silos also tend to lead to inconsistency as the same information will be interpreted in different ways by different risk teams. This disparate relationship between risk teams can lead to the failure to recognize potential correlations between various risks. For example, the recent subprime mortgage crisis that has affected so many banks may have been partially avoided if there had been more coordination and communication between the banks' credit departments and those selling mortgages to people with bad credit. Or if the various regulators, whose function it is to reduce those risks, particularly catastrophic risks, were more forthcoming in sharing information with one another and preferred cooperation to turf protection. Similarly the €6.4 billion ($7 billion) loss at Société Générale was the result of several risk oversights, combining a lack of control on individual traders as well as a failure to implement various checks on the trading systems themselves. Also contributing was a negligence of market risk factors with risk management failing to highlight a number of transactions having no clear purpose or economic value.

1.6.1.3   Tearing down silos

Major risk events rarely result from one risk; rather they commonly involve the accumulation of a number of potential exposures. Consequently, companies need to coordinate better their risk management functions and establish consistent risk reporting mechanisms across their organizations. Applying this discipline to enterprise-wide risk management can be exceptionally difficult given that risk information is often delivered in inconsistent formats. For example, interest rate risk may be reported as a single value at risk (VaR) number, whereas regulatory compliance or operational risk may be expressed through a traffic-light format. This disparity can make it extremely difficult for a CRO, CEO or any senior executive accurately to rank risk exposures. As a result, organizations are now recognizing the need to establish a common framework for reporting risk. This is being undertaken through various initiatives across different industries – ICAS, Solvency II and the Basel II Accord. These initiatives have contributed to the growth of risk and compliance teams. However, the intent of these regulations is not simply to require firms to fulfil their most basic regulatory requirement and to set aside a defined sum of money to cover a list of risk scenarios. Instead, regulators want firms to concentrate on the methodology used to arrive at their risk assessments and to ensure that the risk management process is thoroughly embedded throughout the organization. This requires sound scenario analyses that bring together risk information from all of the various risk silos. It is worthwhile to note that silos do not exist only in the area of risk management. They tend to show up everywhere in organizations where lack of cooperation, competition among units and tunnel vision are allowed to rein unchecked. A notable example of silos is that of the development of separate information systems for the different functional business divisions in an organization, a phenomenon that until the advent and relatively widespread adoption of enterprise-wide computer systems (like ERP, CRM, etc.) caused business untold billions of dollars in losses, wasted and duplicated efforts and lack of coordination within the business. It is high time that risk management adopted the same attitude.

1.6.1.4   Improving audit coordination

Scenario analysis is very much based on the ability to collate and correlate risk information from all over the organization. This includes close coordination not just across the various risk areas, but also with the internal audit teams. This ensures they are more effective and not simply repeating the work of the risk and compliance teams, but rather adding value by rigorously testing this work. Such a task requires using the same common framework as the risk and compliance teams so that information can be seen in the correct context. When this occurs, everyone benefits. Companies are seeing much greater independence and objectivity in the internal audit role. In an increasing number of organizations the internal audit function is no longer confined to existing within a corner of the finance department and has more direct communication with senior management.

1.6.2   Technology's critical role

The use of integrated technology to facilitate the evolution of the three lines of defence is a relatively new development, but will become essential in ensuring coordination across the three lines. Because it has been hard to clarify the different lines of defence and their relationships, it has been difficult to build a business case for a new system and to build the necessary workflow around these different roles. However, the current technology situation, where completely separate legacy systems are used in the business, risk and audit departments, is becoming intolerable and simply contributing to risk. Everyone is aware of the weaknesses in their own systems, but this knowledge does not always translate across the three lines of defence. This leaves most companies with two choices. The first is to design a new all-encompassing system from scratch. The second is to deploy a system that supports common processes and reporting while allowing each function to continue using specialist solutions that suits its own needs. Successful firms will be those that recognize there are different functionalities in these different spaces, but they are all able to communicate with each other in a common language and through common systems. For example, observations can be shared and specific risk issues can then be discussed through an email exchange and summary reports can be automatically sent out to managers.

For internal auditors, a system that supports common processes and reporting improves efficiency and accuracy. The system can enable all lines of defence to establish risk and control libraries, so that where a risk is identified in one office or department, the library can then be reviewed to see if this risk has been recognized and if there are processes in place to manage this risk. Automating risk identification enables companies to take a smarter, more efficient and more global approach to the internal audit function. For business and risk managers, a system that supports common processes makes risk and compliance much simpler. Risk teams have a limited set of resources and must rely on the business to carry out much of the risk management process. This includes conducting risk and control self-assessments, and recording any losses and control breaches where these losses occur. Using a system that supports common processes means that business managers can accurately and efficiently contribute important information, while not being asked to duplicate efforts across risk silos. Risk managers also can then concentrate on the value-added side of their work and their role.

1.6.3   Bringing business into the fold

Beyond simply helping to get the work done, there are far wider benefits to the organization from using systems that support common processes and the principle behind them. For example, the more front-office staff are exposed to the mechanics of the risk management process (rather than being repeatedly petitioned for the same information from multiple parties), the more they are aware of its importance and their role in it.

A couple of decades ago, total quality management was a fashionable concept in many organizations. In some cases, a dedicated management team was assigned to this area, and the rest of the business could assume that quality was no longer their problem, but someone else's. This same misconception applies to risk and compliance, unless all management and employees are kept well informed of such processes and their own active role in them.

Today, it is indeed critically important that everyone realizes that risk is their responsibility. This requires a clear and open line of communication and coordination between three lines of defence: business, risk and compliance, and audit. In order to implement ERM within an organization, the key challenge facing organizations and the CROs is the myriad of risk approaches and systems implemented throughout the modern large institution. Not only is there a huge amount of disparate data to deal with, but the basis on which this data is created and calculated is often different throughout the organization. As a result, it becomes almost impossible to view risks across units, types, countries or business lines.

Another side of the challenge facing CROs is that there are many disparate customers for ERM reporting and analysis. Reports need to be provided to senior business line management, directors and board committees, regulators, auditors, investors, etc. Quite often these customers have different agendas, data requirements, security clearances and format requirements. Often armies of risk analysts are employed within the ERM team whose task is to take information from business and risks systems and manually sort, review and merge this to attempt an overall view of the risk position of the company. This process is very resource and time consuming and extremely prone to error.

In other cases, CROs tackle ERM in a piecemeal fashion. They choose certain risk types or business lines that they feel can be successfully corralled and develop an ERM system to load data concerning those risk types or business lines, normalize that data so that it can be collated and then implement an analytic system to review the enterprise risk within the corral. The aim is to generate a quick win and then expand the framework as methodologies and resources become available. While this approach is a pragmatic one, and derives benefit for the organization, it has one major flaw. If you do not consider the entire picture before designing the approach, it can often be impossible to graft on further types of risk or business line in the future. Even if you manage to make the new addition, the design can fall into the 'I wouldn't have started from here' problem and therefore compromise the entire framework.

What is needed is an approach that implements a general ERM framework from the start that can be utilized as needed by the organization. This framework should cover all risk types and provide support for any business line type or risk measurement type. It should enable an organization to collate data in a standard format without requiring changes to specific lines of business or risk management systems. The 14 chapters of this book provide answers and examples for such a framework using state-of-the-art semantic and analytical technologies.

1.7 Summary

The chapter introduces the concept of risk, defines it and classifies it. We also show the evolution of risk management from none at all to today's heightened awareness of the necessity to deploy enterprise risk management approaches. Risk is now at the core of many applications. For example, Bai and Kenett (2009) propose a risk-based approach to effective testing of web services. Without such testing, we would not be able to use web applications reliably for ordering books or planning a vacation. Kenett *et al.* (2009) present a web-log-based methodology for tracking the usability of web pages. Risks and reliability are closely related. The statistical literature includes many methods and tools in these areas (see Kenett and Zacks, 1998; Hahn and Doganaksoy, 2008). Two additional developments of risks are worth noting. The first one is the introduction of Taleb's concept of black swans. A black swan is a highly improbable event with three principal characteristics: (1) it is unpredictable; (2) it carries a massive impact;

and (3) after the fact, we concoct an explanation that makes it appear less random, and more predictable, than it was (Taleb, 2007). Addressing black swans is a huge challenge for organizations of all size, including governments and not-for-profit initiatives. Another development is the effort to integrate methodologies from quality engineering with risk economics (Kenett and Tapiero, 2009). The many tools used in managing risks seek, de facto, to define and maintain the quality performance of organizations, their products, services and processes. Both risks and quality are therefore relevant to a broad number of fields, each providing a different approach to their measurement, their valuation and their management which are motivated by psychological, operational, business and financial needs and the need to deal with problems that result from the uncertainty and their adverse consequences. Both uncertainty and consequences may be predictable or unpredictable, consequential or not, and express a like or a dislike for the events and consequences induced. Risk and quality are thus intimately related, while at the same time each has, in some specific contexts, its own particularities. When quality is measured by its value added and this value is uncertain or intangible (as is usually the case), uncertainty and risk have an appreciable effect on how we deal, measure and manage quality. In this sense, both risk and quality are measured by 'money'. For example, a consumer may not be able to observe directly and clearly the attributes of a product. And, if and when the consumer does so, this information might not be always fully known, nor be true. Misinformation through false advertising, unfortunate acquisition of faulty products, model defects, etc., have a 'money effect' which is sustained by the parties (consumers and firms) involved. By the same token, poor consumption experience in product and services can have important financial consequences for firms that can be subject to regulatory, political and social pressures, all of which have financial implications. Non-quality, in this sense, is a risk that firms assess, that firms seek to value and price, and that firms manage to profit and avoid loss. Quality and risk are thus consequential and intimately related. The level of delivered quality induces a risk while risk management embeds tools used to define and manage quality. Finally, both have a direct effect on value added and are a function of the presumed attitudes towards risk and the demands for quality by consumers or the parties involved in an exchange where it is quality or risk.

This introductory chapter lays the groundwork for the whole book that will move us from the general view of risk to specific areas of operational risk. In the following chapters the reader will be presented with the latest techniques for operational risk management coming out of active projects and research dedicated to the reduction of the consequences of operational risk in today's highly complex, fast-moving enterprises. Many examples in the book are derived from work carried out within the MUSING project (MUSING, 2006). The next chapter provides an introduction to operational risk management and the successive 12 chapters cover advanced methods for analysing semantic data, combining qualitative and quantitative information and putting integrated risk approaches at work, and benefiting from them. Details on operational risk ontologies and data mining techniques for unstructured data and various applications are presented, including their implication to intelligent regulatory compliance and the analysis of near misses and incidents.

The overall objective of the book is to pave the way for next generation operational risk methodologies and tools.

References

Alexander, C.O. (1998) *The Handbook of Risk Management and Analysis*, John Wiley & Sons, Inc., New York.

Ayyub, B.M. (2003) *Risk Analysis in Engineering and Economics*, Chapman & Hall/CRC Press, Boca Raton, FL.

Bai, X. and Kenett, R.S. (2009) Risk-Based Adaptive Group Testing of Web Services, *Proceedings of the Computer Software and Applications Conference (COMPSAC'09)*, Seattle, USA.

Chorafas, D.N. (2004) *Operational Risk Control with Basel II: Basic Principles and Capital Requirements*, Elsevier, Amsterdam.

Davies, J.C. (Editor) (1996) *Comparing Environmental Risks: Tools for Setting Government Priorities*, Resources for the Future, Washington, DC.

Doherty, N.A. (2000) *Integrated Risk Management: Techniques and Strategies for Managing Corporate Risk*, McGraw-Hill, New York.

Dowd, K. (1998) *Beyond Value at Risk: The New Science of Risk Management*, John Wiley &, Ltd, Chichester.

Eidinow, E. (2007) *Oracles, Curses & Risk Among the Ancient Greeks*, Oxford University Press, Oxford.

Embrecht, P., Kluppelberg, C. and Mikosch, T. (1997) *Modelling External Events*, Springer-Verlag, Berlin.

Engelmann, B. and Rauhmeier, R. (2006) *The Basel II Risk Parameters*, Springer, Berlin–Heidelberg, Germany.

Finkel, A.M. and Golding, D. (1994) *Worst Things First? The Debate over Risk-Based National Environmental Priorities*, Resources for the Future, Washington, DC.

Guess, F. (2000) Improving Information Quality and Information Technology Systems in the 21st Century, *International Conference on Statistics in the 21st Century*, Orino, ME.

Hahn, G. and Doganaksoy, N. (2008) *The Role of Statistics in Business and Industry*, Wiley Series in Probability and Statistics, John Wiley & Sons,Inc., Hoboken, NJ.

Haimes, Y.Y. (2009) *Risk Modeling, Assessment and Management*, third edition, John Wiley & Sons, Inc., Hoboken, NJ.

Jorion, P. (1997) *Value at Risk: The New Benchmark for Controlling Market Risk*, McGraw-Hill, Chicago.

Kenett, R.S. (2008) From Data to Information to Knowledge, *Six Sigma Forum Magazine*, pp. 32–33.

Kenett, R.S. (2009) Discussion of Post-Financial Meltdown: What Do the Services Industries Need From Us Now?, *Applied Stochastic Models in Business and Industry*, 25, pp. 527–531.

Kenett, R.S. and Raphaeli, O. (2008) Multivariate Methods in Enterprise System Implementation, Risk Management and Change Management, *International Journal of Risk Assessment and Management*, 9, 3, pp. 258–276 (2008).

Kenett, R.S. and Salini, S. (2008) Relative Linkage Disequilibrium Applications to Aircraft Accidents and Operational Risks, *Transactions on Machine Learning and Data Mining*, 1, 2, pp. 83–96.

Kenett, R.S. and Shmueli, G. (2009) On Information Quality, University of Maryland, School of Business Working Paper RHS 06-100, http://ssrn.com/abstract=1464444 (accessed 21 May 2010).

Kenett, R.S. and Tapiero, C. (2009) Quality, Risk and the Taleb Quadrants, *Risk and Decision Analysis*, 4, 1, pp. 231–246.

Kenett, R.S. and Zacks, S. (1998) *Modern Industrial Statistics: Design and Control of Quality and Reliability*, Duxbury Press, San Francisco.

Kenett, R.S., de Frenne, A., Tort-Martorell, X. and McCollin, C. (2008) The Statistical Efficiency Conjecture, in *Statistical Practice in Business and Industry*, Coleman, S., Greenfield, T., Stewardson, D. and Montgomery, D. (Editors), John Wiley &Sons, Ltd, Chichester.

Kenett, R.S., Harel, A. and Ruggerri, F. (2009) Controlling the Usability of Web Services, *International Journal of Software Engineering and Knowledge Engineering*, 19, 5, pp. 627–651.

Knight, F.H. (1921) *Risk, Uncertainty and Profit*, Hart, Schaffner and Marx (Houghton Mifflin, Boston, 1964).

MUSING (2006) IST- FP6 27097, http://www.musing.eu (accessed 21 May 2010).

Panjer, H. (2006) *Operational Risks: Modelling Analytics*, John Wiley & Sons, Inc., Hoboken, NJ.

Redman, T. (2007) Statistics in Data and Information Quality, in *Encyclopedia of Statistics in Quality and Reliability*, Ruggeri, F., Kenett, R.S. and Faltin, F. (Editors in chief), John Wiley & Sons, Ltd, Chichester.

Taleb, N.N. (2007) *The Black Swan: The impact of the highly improbable*, Random House, New York.

Tapiero, C. (2004) *Risk and Financial Management: Mathematical and Computational Methods*, John Wiley & Sons, Inc., Hoboken, NJ.

Van den Brink, G. (2002) *Operational Risk: The New Challenge for Banks*, Palgrave, New York.

2

Operational risk management: an overview

Yossi Raanan, Ron S. Kenett and Richard Pike

2.1 Introduction

Operational risk management is a somewhat new discipline. While financial risks were recognized long ago, they are in fact part of everyday life and not just a business issue; operational risks and their management have been misdiagnosed frequently as human error, machine malfunction, accidents and so on. Often these risks were treated as disconnected episodes of random events, and thus were not managed. With the advancement of computerized systems came the recognition that operational mishaps and accidents have an effect, sometimes a very considerable one, and that they must be brought under control. Today, operational risk management is gaining importance within businesses for a variety of reasons. One of them is the regulatory demand to do so in important sectors of the economy like banking (Basel II, 2006), insurance (Solvency II, 2009) and the pharmaceutical industry (ICH, 2006). Another is the recognition that since operations are something that the business can control completely or almost completely, it ought also to manage the risk associated with these operations so that the controls are more satisfactory for the various stakeholders in the business. This chapter provides an overview of operational risk management (OpR) and enterprise risk management (ERM) as background material for the following chapters of the book.

Operational Risk Management: A Practical Approach to Intelligent Data Analysis Edited by Ron S. Kenett and Yossi Raanan © 2011 John Wiley & Sons, Ltd

2.2 Definitions of operational risk management

Operational risk has a number of definitions which differ mainly in details and emphasis. Although the proper definition of operational risk has often been the subject of past heated debate (International Association of Financial Engineers, 2010), there is general agreement among risk professionals that the definition should, at a minimum, include breakdowns or failures relating to people, internal processes, technology or the consequences of external events. The Bank for International Settlements, the organization responsible for the Basel II Accord regulating risk management in financial institutions, defines operational risk as follows:

> Operational risk is defined as the risk of loss resulting from inadequate or failed internal processes, people and systems or from external events. This definition includes legal risk, but excludes strategic and reputational risk. Legal risk includes, but is not limited to, exposure to fines, penalties, or punitive damages resulting from supervisory actions, as well as private settlements.
>
> (Basel II, 2006)

It is this latter definition that will be used here. In layman's terms, operational risk covers unwanted results brought about by people not following standard operational procedures, by systems, including computer-based systems, or by external events.

In the Basel II definition, 'inadequate or failed internal processes' encompass not only processes that are not suitable for their purpose, but also processes that failed to provide the intended result. These, of course, are not the same. Processes may become unsuitable for their purpose due to external events, like a change in the business environment over which the business has no control. Such change might have been so recent that the business or organization did not have the time to adjust itself. Failed processes, on the other hand, mean that the organization has fallen short in their design, implementation or control. Once we include internal auditing as one of the important business processes, it is seen that internal fraud and embezzlements are part of the definition.

The 'people' part covers both the case of human error or misunderstanding and the case of intentional actions by people – whether with intent to cause harm, defraud or cheat, or just innocently cutting corners, avoiding bureaucratic red tape or deciding that they know a better way of executing a certain action. 'Systems' covers everything from a simple printer or fax machine to the largest, most complicated and complex computer system, spread over many rooms, connecting many users and many other stakeholders located in every corner of the globe. Last in this shortlist of categories of operational risk is 'external events'. This innocently looking phrase covers a lot of possible causes for undesired outcomes – from hackers trying to disrupt computer systems, through labour strikes, to terrorist attacks, fires or floods.

Operational risks abound in every sector of the economy and in every human endeavour. Operational risks are found in the health sector, in the transportation sector, in the energy industry, in banking, in education and, indeed, in all activities. Some sectors, because of enhanced sensitivity to risks or because of government regulations, have implemented advanced processes for identifying the risks particular to their activities. However, operational risks exist when any activity occurs, whether we manage them or not. This recognition is beginning to reach the awareness of many management teams in a wide variety of activities (Doebli *et al.*, 2003).

An example where operational risks are recognized as a source for large potential losses can be found in the report by the Foreign Exchange Committee (2004) that encourages best practices for the mitigation of operational risks in foreign exchange services. A detailed discussion of risk management in this industry, including an application of Bayesian networks used later in this book, can be found in Adusei-Poku (2005).

On 14 March 2010, the *Sunday Times* published a summary of a 2200-page report investigating the crash of Lehman Brothers on Wall Street described in Chapter 1 (*Sunday Times*, 2010). The report stated that, on May 2008, a senior vice president of Lehman Brothers wrote a memo to senior management with several allegations, all of which proved right. He claimed that Lehman had 'tens of billion of dollars of unsubstantiated balances, which may or may not be 'bad' or non-performing assets or real liabilities', and he was worried that the bank had failed to value tens of billion of dollars of assets in a 'fully realistic or reasonable way' and did not have staff and systems in place to cope with its rapid growth.

Lehman's auditors, Ernst & Young, were worried but did not react effectively. Time was not on Ernst & Young or Lehman Brother's side. By September, the 158-year-old bank was bust, thousands of people had lost their jobs and the world's economy was pitched into a black hole. The court-appointed bankruptcy examiner found Lehman used accounting jiggery-pokery to inflate the value of toxic real-estate assets it held, and chose to 'disregard or overrule the firm's risk controls on a regular basis'. His most juicy finding was Repo 105, which the report alleges was used to manipulate the balance sheet to give the short-term appearance of reducing assets and risk. Not since Chewco and Raptor – Enron's 'off balance sheet vehicles' – has an accounting ruse been so costly.

These events are all examples of operational risks.

In summary, operational risks include most of what can cause an organization harm, that is foreseeable and, to a very large extent, avoidable – if not the events themselves, then at least their impact on the organization. It is quite plain that once we recognize the operational risks that face our enterprise, we can mitigate them. It is important to understand that a risk, once identified, is no longer a risk – it is a management issue. OpR is the collection of tools, procedures, assets and managerial approach that are all aimed together at one goal: to understand the operational risks facing the enterprise, to decide how to deal with them and to manage this process effectively and efficiently. It should be noted that the idea of OpR is, in some sense, a circular problem. The processes and systems used for managing operational risks are all subject, themselves, to the same pitfalls that may cause systems and people to malfunction in other parts of the organization. It is hoped, however, that once OpR is adopted as a basic approach of management, the OpR system itself will be subjected to the same testing, screening and control that every other aspect of the operation is subjected to.

2.3   Operational risk management techniques

2.3.1   Risk identification

In order to manage and control risk effectively, management need a clear and detailed picture of the risk and control environment in which they operate. Without this knowledge, appropriate action cannot be taken to deal with rising problems. For this purpose, risks must be identified. This includes the sources, the events and the consequences of the risks. For this and other risk-related definitions, see also ISO 73 (2009).

Every organization has generic activities, processes and risks which apply to all business areas within the organization. Risk descriptions and definitions should be stored in one repository to allow organizations to manage and monitor them as efficiently as possible. This approach creates a consolidated, organization-wide view of risk, regardless of language, currency, aggregation hierarchy or local regulatory interpretations.

This consolidated view allows the organization to monitor risk at a business unit level. However, it is integral for each business unit to identify and monitor its local risks, as the risks may be unique to that business unit. In any case, a business unit is responsible for its results and thus must identify the risks it faces. In order to do this effectively, risks must be identified. Notwithstanding risks that are common knowledge, like fire, earthquakes and floods, they must also be included in the final list. All other risks, specific to the enterprise, must be identified by using a methodology designed to discover possible risks. This is a critical step, since management cannot be expected to control risks they are unaware of. There are a number of ways of identifying risks, including:

- Using event logs to sift the risks included in them.

- Culling expert opinions as to what may go wrong in the enterprise.

- Simulating business processes and creating a list of undesirable results.

- Systematically going through every business process used in the enterprise and finding out what may go wrong.

- Using databanks of risk events that materialized in similar businesses, in order to learn from their experience.

Some of these methods produce only a list of risks, while others may produce some ideas, more or less accurate, depending on the particular realization of the frequency of these risk events actually happening. This frequency is used to calculate the expected potential damage that may become associated with a particular event and, consequently, for setting the priorities of treating various contingencies.

Organizations ensure consistency in risk identification in two ways:

1. Risk identification is achieved via a centralized library of risks. This library covers generic risks that exist throughout the organization and associates the risks with the organization's business activities. When a business unit attempts to define its local risks and build its own risk list, it does so by considering a risk library. The library itself is typically created by using an industry list as an initial seed, and then augmented by collecting risk lists from every business unit, or it may be created by aggregating the risks identified by each business unit. In either case, this process must be repeated until it converges to a comprehensive list.

2. Identification consistency is further aided by employing a classification model covering both risks and controls. Using this model each risk in the library has an assigned risk classification that can be based on regulatory definitions, and each associated control also has a control classification. The key benefits of classification are that it allows organizations to identify common risks and control themes.

Once risks have been identified, control must be put in place to mitigate those risks. Controls can be defined as processes, equipment or other methods, including knowledge/skills and organization design, that have a specific purpose of mitigating risk. Controls should be identified and updated on a regular basis.

Controls should be:

- Directly related to a risk or a class of risks (not a sweeping statement of good practice).

- Tangible and normally capable of being evidenced.

- Precise and clear in terms of what specific action is required to implement the control.

The process of risk identification should be repeated at regular intervals. This is because risks change, the nature of the business evolves, the regulatory climate (sometimes defining which risks must be controlled) changes, the employees are rotated or replaced, new technologies appear and old technologies are retired. Thus, the risk landscape constantly evolves and, with it, the risks.

2.3.2   Control assurance

A control assurance process aims to provide assurance throughout the business that controls are being operated. It is generally implemented in highly 'control focused' areas of the business where management and compliance require affirmation that controls are being effectively operated.

Control assurance reporting is defined as the reporting of the actual status of a control's performance. This is fundamentally different from the risk and control assessment process discussed in Section 2.3.4, which is concerned with assessing and validating the risk and control environment. Control assurance is a core component of the risk management framework and is used to:

- Establish basic transparency and reporting obligations.

- Establish where 'control issues' occur and ensure that the relevant management actions are taken.

- Highlight insufficiently controlled areas.

- Highlight areas of 'control underperformance'.

- Provide detailed control reporting to various levels of management.

Control assurance is not necessarily undertaken by every area in the business; it is more noticeably present in the areas of the business that require assurance that controls are being effectively operated.

Control assurance is generally performed on a periodic basis, typically monthly or quarterly. Each business unit typically nominates someone to ensure that control assurance reporting is carried out. This does not mean that this is the only person who has controls to operate; rather this person ensures that all controls have been operated by the relevant person in the area for which he/she is responsible.

Business units, in conjunction with appropriate risk management personnel, should define all of the controls within their responsibility. From this, the shortlist of controls to be included in the control assurance process is developed. This shortlist should consider:

- The impact and likelihood of the risk mitigated by the control.

- The effectiveness and importance of the control.

- The frequency of the control operation.

- The regulatory relevance of the control.

- The cost/performance ratio of developing and implementing the control.

The OpR function monitors the control shortlists in conjunction with business units to ensure their appropriateness and adequacy.

2.3.3   Risk event capture

Risk event capture is the process of collecting and analysing risk event data. An operational risk event, as previously defined, can result in:

- An actual financial loss of a defined amount being incurred – a loss.

- An actual financial profit of a defined amount being incurred – a profit.

- A situation where no money was actually lost but could have been were it not for the operation of a control – a near miss.

- A situation where damage is caused to equipment and to people.

When analysing risk events, it should be possible to identify:

- The controls which failed or the absence of controls that allowed the event to occur.

- The consequence of the event in terms of actual financial loss or profit.

- The correlations between risks – as a financial loss is often the result of more than one risk co-occurring.

Although collecting risk event data is in many cases an external regulatory requirement, it is also beneficial to an organization in that it:

- Provides an understanding of all risk events occurring across the organization.

- Provides quantifiable historical data which the organization can use as input into modelling tools.

- Promotes transparent and effective management of risk events and minimizes negative effects.

- Promotes root cause analysis which can be used to drive improvement actions.

- Reinforces accountability for managing risk within the business.

- Provides an independent source of information which can be used to challenge risk and control assessment data.

The degree of cooperation of front-line workers with the reporting requirements varies and is not uniform – not across industries and not even across a particular organization. As Adler-Milstein *et al.* (2009) show, workers are more likely to report operational failures that carry financial or legal risks.

2.3.4   Risk and control assessments

The management of risks and their associated controls is fundamental to successful risk management. Any risk and control assessment (RCA) process should be structured and consistent to allow for the qualitative assessment of the validity of key business risks and their controls. This is fundamentally different from control assurance which is concerned with providing assurance that controls are being effectively operated.

RCA is a core component of the risk management framework and is used to:

- Identify the key risks to the business.

- Assess the risks in terms of their overall significance for the business based on the judgement of business management.

- Establish areas where control coverage is inadequate.

- Drive improvement actions for those risks which are assessed as outside agreed threshold limits for risk.

- Provide consistent information on the risk and control environment which can be aggregated and reported to senior management to better help in making more informed decisions.

RCA is performed in different areas of the organization, referred to as assessment points. These are identified by the relevant business unit owners. RCA is generally performed on a periodic basis, typically monthly or quarterly. The duration of each assessment is variable and will depend on the number of risks and controls to be assessed. Both business unit owners and members of the risk management team will be involved in each RCA.

RCA is normally a three-step process which allows the business to identify, assess and manage risk:

1. The identification step (which takes place outside of any system) results in a list of the key risks to be included in the assessment.

2. The assessment step allows the business to rank the risks identified in terms of significance to the business and assess the validity of their scoring. This step will include an approval of the assessment.

3. The management step is primarily involved with ensuring improvement actions raised as a result of risks being outside agreed limits are followed up and compiling reporting information.

One of the goals of this activity is to be able to predict the risks facing the organization, so that the priorities for handling them can be properly decided. That is, the goal is to be able to manage the operational risk and bring its size to that level which the organization can tolerate. It is not just about bookkeeping and clerical record keeping, done in order to demonstrate diligence. As Neil *et al.* (2005) note, 'Risk prediction is inextricably entwined with good management practice and [that] measurement of risk can meaningfully be done only if the effectiveness of risk and controls processes is regularly assessed.'

2.3.5   Key risk indicators

Key risk indicators, or KRIs, are metrics taken from the operations of a business unit, which are monitored closely in order to enable an immediate response by the risk managers to evolving risks. This concept of 'Key X indicators' is not new, nor is it particular to risk management. Its more familiar form is KPI, where P stands for Performance. The basic idea behind these two acronyms is quite similar. Indicators – for risk or for performance – may be quite numerous within a given enterprise. For an industrial firm risk indicators may include:

- Number of defective items produced – in each production line.

- Percentage of defective items produced – in each production line.

- Change – daily, weekly, monthly, etc. – in the number of defective items produced in each production line.

- Number of items returned as defective for each product (again, this may be expressed in numbers, percentages or monetary value).

- Number of maintenance calls for each production line – absolute or per unit of time.

- Number of accidents on the production lines.

- Number of unplanned stoppages of each production line.

For achieving comprehensive OpR in an enterprise, we add to the KPIs listed above operational risk indicators associated with other divisions of the enterprise – finance, marketing, human resources and computer operations. So, it is evident that the number of risk indicators in a given enterprise may be very large, thus making it very difficult to track, monitor and control. Therefore, a select few risk indicators are chosen to serve as a warning mechanism for the enterprise. These may be simple risk indicators like 'number of computer crashes in a week', or 'number of communication breakdowns in a day', or 'costs of unscheduled repairs incurred in the computer centre during a prescribed period of time'. Alternatively, they may be compound indicators, artificial in a sense, made up of direct risk indicators for a given area of activity to create a representative indicator for that activity in such a way that changes in this compound indicator will warn the risk management officer of approaching difficulties.

The KRIs are lagging or leading indicators of the risks facing the enterprise. The way to create them changes from one organization to another, and their construction expresses such attributes as the level of importance that the organization attaches to each of its activities, the regulatory climate under which the organization operates and the organization's appetite for risk. Consequently, two similar organizations serving the same markets may have quite different KRIs. The list of possible KRIs is so long – when compiled from all possible sources – that libraries of KRIs have been set up and some can only be accessed under a subscription agreement – see, for example, KRIL (2010). The actual definition of a particular organization's KRIs requires usually a project targeted at this goal that is usually undertaken as part of an overall OpR approach. For more on KPIs and KRIs see Ograjenšek and Kenett (2008) and Kenett and Baker (2010). A study by Gartner positioning OpR software products is available in McKibben and Furlonger (2008).

2.3.6   Issues and action management

The management of issues and their associated actions is fundamental to successful OpR. The issues and actions management process should provide a standardized mechanism for identifying, prioritizing, classifying, escalating and reporting issues throughout the company.

The collection of issues and actions information allows the business to adopt a proactive approach to OpR and allows for swift reactions to changes in the business environment.

Issues and actions management is a core component of the risk management framework and is used to:

- Support the evaluation of risk likelihood and control effectiveness during the RCA process.

- Highlight control failures or uncontrolled risks during the control assurance process.

- Highlight events resulting in significant financial loss.

Guiding principles state that issues should generally originate from:

- Control improvements.

- Control weaknesses.

- Compliance gaps/concerns.

- Audit recommendations – both financial audit and risk audit.

- Risk event reports.

- Quality defects.

The issue management process should:

- Capture issues related to the RCA and control assurance processes, risk events, internal audits and compliance audits.

- Support the creation of issues on an ad hoc basis.

- Allow for the creation of actions and assign responsibilities and target completion dates for the same.

- Monitor the satisfactory completion of issues and actions.

- Provide reports to support the issue management and action planning process.

2.3.7   Risk mitigation

Risk mitigation is an action, consciously taken by management, to counteract, in advance, the effects on the business of risk events materializing. The risk mitigation strategies for operational risks fall into the same four general categories of risk mitigation used for managing risks of all types. These are:

- Avoid the risk.

- Accept the risk.

- Transfer the risk.

- Reduce the risk.

Avoiding the risk means not taking the action that may generate it. With operational risk, that means not performing the operation. Accepting the risk means that the organization, while well aware of the risk, decides to go ahead and perform the operation that may end in the risk event occurring, and to suffer the consequences of that occurrence. Transferring the risk may be accomplished by a number of methods. The most familiar one is to insure the business against the occurrence of that risk event. This way, the risk is transferred to the insurance company and a probabilistic loss event (the risk actually occurring and causing damage) is substituted by a deterministic, known loss – the insurance premium. Another way of transferring the risk is to subcontract the work that entails the risk, thereby causing some other business to assume the risk. Finally, reducing the risk means taking steps to lower either the probability of the risk event happening or the amount of damage that will be caused if it does occur. It is possible to act on these two distributions simultaneously, thereby achieving a lower overall risk.

Risk mitigation is an important part of risk management in general and operational risk is no exception. In some sense, the area of OpR that is restricted to the management of information and communications technology (ICT) operations has been concerned for quite some time with disaster recovery planning (DRP), which is a detailed plan for continued ICT operations in case a disastrous event happens. However, DRP deals with major disruptions of ICT operations in the enterprise, while risk management deals with all types of risks, large and small. Recently, this area of risk mitigation has been extended to the whole business and the area of business continuity management deals with the ways and means to keep a business going even after a major catastrophe strikes.

2.4   Operational risk statistical models

Operational risks are characterized by two statistical measures related to risk events: their severity and their frequency (Cruz, 2002). A common approach to model the frequency and the severity is to apply parametric probability distribution functions. For severity, the normal and lognormal distributions are often applied. Other distributions used to model the severity are: inverse normal, exponential, Weibull, gamma and beta. For details on these distributions see Kenett and Zacks (1998).

On the other hand, in order to model the frequency of specific operational risk events, two main classes are used: ordinary (Poisson, geometric, binomial) and zero-truncated distributions.

The most common goodness-of-fit test for determining if a certain distribution is appropriate for modelling the frequency of events in a specific data set is the chi-square test. The formal test for testing the choice made for a severity distribution is instead the Kolmogorov–Smirnov test and related measures of interest (see Kenett and Zacks, 1998).

Having estimated, separately, both the severity and the frequency distributions, in operational risk measurement we need to combine them into one aggregated loss distribution that allows us to predict operational losses with an appropriate degree of confidence. It is usually assumed that the random variables that describe severity and frequency are stochastically independent. Formally, the explicit formula of the distribution function of the aggregated losses, in most cases, is often not analytically explicit. One popular practical solution is to apply a Monte Carlo simulation (see Figure 2.1).

On the basis of the convolution obtained following a Monte Carlo simulation, operational risk measurement can be obtained as a summary measures, such as the 99.9th percentile of the annual loss distribution, also called value at risk (VaR). In operational risk the distribution of a financial loss is obtained by multiplying the frequency distribution by the severity distribution. These considerations motivate the use of the geometric mean of risk measures, when aggregating risks over different units. The use of the geometric mean is a necessary condition for preserving stochastic dominance when aggregating distribution functions.

Cause and effect models have also been used extensively in operational risk modelling. Specifically Bayesian methods, including Bayesian networks, have been proposed for modelling the linkage between events and their probabilities. For more on these methods see Alexander (2000, 2003), Giudici and Billota (2004), Cornalba and Giudici (2004), Bonafede and Giudici (2007), Fenton and Neil (2007), Ben Gal (2007), Dalla Valle et al. (2008), Figini et al. (2010), Kenett (2007) and Chapters 7 and 8 in this book. These and the next chapters include examples from the MUSING project (MUSING, 2006). The next section presents a short overview of classical operational risk measurement techniques.

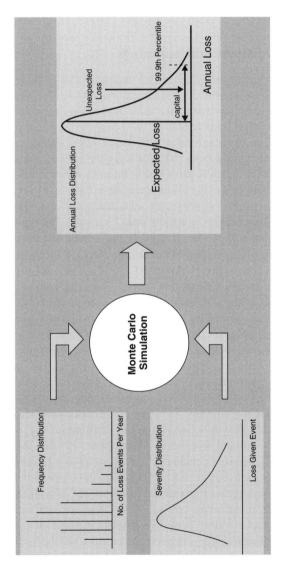

Figure 2.1   Monte Carlo convolution of frequency and severity.

2.5   Operational risk measurement techniques

In order to be able to assess and manage risk, it must be measured. It is impossible to manage anything that is not measured, risk being a prime example of this approach. In this section we introduce three operational risk measurement techniques: the loss distribution approach, scenario analysis and balanced scorecards.

2.5.1   The loss distribution approach

The loss distribution approach (LDA) is a measurement technique that is particularly suitable for banks and other financial institutions. It aims at calculating the VaR, which is a monetary value that these institutions need in order to assign adequate capital, as far as their regulators are concerned, against operational risk (see Figure 2.1). This expected value may be of lesser interest for businesses that have a different approach to risk, for example if they view small losses, bounded above by a periodically changeable limit, as either negligible or part of the cost of doing business. On the other hand, these businesses insure themselves against losses that surpass another dynamically changed amount and consequently implement mitigation strategies to handle only losses that fall between these two bounds. This optional mitigation strategy is not available to banks and many other financial institutions for they function, in effect, as their own insurers and therefore must have a more precise knowledge of the risks, not just some bounds and frequencies. As an example of this type of risk management behaviour one may look at supermarkets and large food sellers in general that have become accustomed, albeit unwillingly, to losses stemming from employee theft – a definite operational risk. Many consider this theft-produced loss a part of doing business as long as it does not rise above a certain level, determined individually by each chain or food store, and take out a specific policy with an insurance company against larger thefts.

The LDA, which is used extensively in calculating the capital requirements a financial institution has to meet to cover credit risks, is a statistically based method that estimates two functions involved with risk – the occurrence frequency and the loss amount frequency. From these two distributions, the distribution of the VaR may be computed. For financial institutions, the VaR has to be calculated for each business line (Basel II, 2006), and then a total VaR is calculated by summing the individual business line VaRs multiplied by their weight in the bank's outstanding credits. While this measuring method is complex to implement and requires extensive databases, some of them external to the bank, and is computationally intensive, there are a number of approaches for financial institutions to calculate it (see e.g. Frachot *et al.*, 2001; Tapiero, 2004; Shevchenko, 2009). The effort and investments involved may be worthwhile only for large banks, since it can lead to a significantly smaller capital allocation for operational risk, thus freeing a highly valuable resource for the bank.

For operational risk in other types of business, such a very fine breakdown of events and their consequences may not be required, for a number of reasons.

First, the operational risk is seldom, if ever, related to a business line. Second, operational risk events are frequently the result of more than one causing factor in the wrong range and thus attributing the risk to one of them or distributing it among them will be highly imprecise, to say the least. Third, the costs of implementing such a measurement system may prove prohibitive for a business that is capable of getting insurance against these losses for a small fraction of that cost. A method similar to the LDA is demonstrated for a process that is part of OpR in banks in Chapter 10 describing the near miss/opportunity loss in banks.

2.5.2 Scenarios

Scenarios are used in many areas where the prospects of having accurate predictions are slim or where there are no analytical tools available to produce such predictions at all. They are frequently used for strategic planning in order to discover, as realistically as feasible, what would be a suitable reaction by the business to a wide range of possible developments of many variables that affect the business, in various combinations. Scenarios range from an extension of current reality into the foreseeable future to extreme changes in the business's environment, status, capabilities and associations. Scenarios are used in operational risk measurement in a number of cases. The first case involves an organization that wishes to engage in OpR, but lacks the requisite risk event repository from which to calculate – or even simply summarize – the results of the various risks. That is the most usual case, and it is frequently used because it takes a long time from the initiation of a risk management activity to the time when the organization has a workable repository with enough risk events that materialized. Thus, organizations use the scenario technique in order to shorten the time to the implementation of a risk management approach with the proper mitigation strategies. The second case involves a significant change in the environment that the business operates in. Usually it is a change in the external environment: new regulatory demands, radically changed economic environment, new technologies being brought rapidly to bear on the economic segment the business operates in, and so on. Occasionally, it may be a drastic reorganization of the business, such as a merger of different units into a single one, or a merger with another business or an acquisition of a business and the attempt to assimilate it successfully into the business.

The scenarios technique involves a team, familiar with the business processes being studied, devising possible business scenarios – and trying to see what the reaction of the business might be, and what might go wrong. Doing this systematically, step by step, and covering all possible areas (technology, people, processes, etc.) that might be affected by the scenario, results in a list of potential risk events that are latent within the business process under study. This method is then applied to every business process used in the business until a complete list of latent risk events is compiled. This list is then analysed, categorized and stored as a virtual risk event repository. Then, a measure may be computed for variables that are of interest, including the VaR involved with each risk event. If some data is available that describes the frequency of executing a particular business process, estimates of expected losses can be computed. Mitigation strategies are then devised for each risk event, and the implementation of OpR continues from this point onward.

The benefits of this technique are:

1. It is not dependent on an existing repository of risk events.

2. Even if a risk event repository exists in the business, this technique may prepare the business for risk events that have not yet been registered in the repository – for the simple reason that they had not occurred or that they had occurred prior to the repository being established – but these risks are nevertheless worth considering and preparing mitigation strategies for them.

3. It may be done in a relatively short period of time, eliminating the need for waiting for a significant accumulation of risk events in the risk repository.

4. It may be used in addition to using the risk repository.

The drawbacks of this technique are:

1. It is based on a complete mapping of all business processes in the business. Leaving out a few business processes may make the whole effort not useful since significant portions of the business activity may be left uncovered.

2. It usually requires a large team. The team usually includes people from the risk management office, from the industrial engineering unit and from the operation of the business itself. The core people, like the risk managers and the industrial engineers, may form the central, fixed part of the team, but the people familiar with the various business processes will have to change with each area of activity covered.

3. Lacking any significant history of risk events, it requires a very determined management to undertake such an extensive and expensive activity.

All things considered, it is a good technique, though usually the lack of complete mapping of all business processes prevents it from being very effective. On the other hand, this mapping – a requisite for this technique – may be a very substantial side benefit of this operation and, indeed, it may be a sufficient benefit in and of itself so as to justify the whole process.

2.5.3   Balanced scorecards

Scorecards were made famous in the business world by Norton and Kaplan in the early 1990s (Kaplan and Norton, 1992, 1993, 1996; see also Organjenšek and Kenett, 2008). Since that time, the notion has caught on and today the balanced scorecard (BSC) is widely used in businesses in all disciplines. For an application to organizations developing systems and software see Kenett and Baker (2010). In short, the basic concept of the scorecards is, as the name implies, to compute a score for the measured phenomena and to act upon its changing values. The concept of an operational risk scorecard is the same as that of the general scorecard, except that in this case it is much more specialized and concerns only operational risks in the business. Whereas in the classic BSC the scores represent the performance in the financial, customer, internal processes and learning and growth facets of the business (although many variations exist), in the operational risk scorecard the measured aspects may be technology, human factors and external factors affecting the business operations. This division is by no means unique, and many other divisions may be used. For example, a bank trying to comply fully with the Basel II recommendations may concentrate more heavily on the ICT part of the operations when handling operational risk, and subdivide this score into finer categories – hardware, software, communications, security and interface. Similar subdivisions may be tried out in other areas representing operational risk.

When the complete classification and categorization of all operational risks are completed, weights are assigned to the elements within each category and then a risk score may be computed for each category by providing the values of the individual risks of the elements. The resulting score must be updated frequently to be of value to the organization.

As a final note, it is worthwhile to consider a combined risk indicator, composed of the individual risk categories managed by the organization, which is added to its overall scorecard, thus providing management not only with performance indicators in the classic BSC, but also with an indication of the risk level at which the organization is operating while achieving the business-related indicators.

2.6   Summary

This chapter introduces the basic building blocks of operational risk management, starting from the basic definition of operational risk, through the steps of identifying, classifying, controlling and managing risks. The following chapters, organized in three parts, provide an in-depth analysis of the various ways and means by which operational risk are handled. We briefly describe these three parts.

Part II: Data for Operational Risk Management and its Handling

Operational risk management relies on diverse data sources, and the handling and management of this data requires novel approaches, methods and implementations. This part is devoted to these concepts and their practical applications. The applications are based on case studies that provide practical, real examples for the practitioners of operational risk management. The chapters included in Part II are:

Chapter 3:   Ontology-based modelling and reasoning in operational risks
Chapter 4:   Semantic analysis of textual input
Chapter 5:   A case study of ETL for operational risks
Chapter 6:   Risk-based testing of web services

Part III: Operational Risks Analytics

The data described in Part II requires specialized analytics in order to become information and in order for that information to be turned, in a subsequent phase of its analysis, into knowledge. These analytical methods are described in the following chapters:

Chapter 7:   Scoring models for operational risks
Chapter 8:   Bayesian merging and calibration for operational risks
Chapter 9:   Measures of association applied to operational risks

Part IV: Operational Risk Management Applications and Integration with other Disciplines

Operational risk management is not a stand-alone management discipline. This part of the book demonstrates how operational risk management relates to other management issues and intelligent regulatory compliance. The chapters in this part consist of:

Chapter 10:   Operational risk management beyond AMA: new ways to
                    quantify non-recorded losses
Chapter 11:   Combining operational risks in financial risk assessment scores
Chapter 12:   Intelligent regulatory compliance
Chapter 13:   Democratization of enterprise risk management
Chapter 14:   Operational risks, quality, accidents and incidents

The book presents state-of-the-art methods and technology and concrete implementation examples. Our main objective is to push forward the operational risk management envelope in order to improve the handling and prevention of risks. We hope that this work will contribute, in some way, to organizations which are motivated to improve their operational risk management practices and methods with modern technology. The potential benefits of such improvements are immense.

References

Adler-Milstein, J., Singer, S.J. and Toffel, M.W. (2009) Operational Failures and Problem Solving: An Empirical Study of Incident Reporting, Harvard Business School Technology and Operations Management Unit, Working Paper No. 10-017. http://ssrn.com/abstract=1462730 (accessed 21 May 2010).

Adusei-Poku, K. (2005) Operational Risk Management – Implementing a Bayesian Network for Foreign Exchange and Money Market Settlement, PhD dissertation, Faculty of Economics and Business Administration of the University of Gottingen.

Alexander, C. (2000) Bayesian Methods for Measuring Operational Risk, http://ssrn.com /abstract=248148 (accessed 21 May 2010).

Alexander, C. (2003) *Operational Risk: Regulation, Analysis and Management*, Financial Times/Prentice Hall, London.

Basel Committee on Banking Supervision (2006) Basel II: International Convergence of Capital Measurement and Capital Standards: A Revised Framework – Comprehensive Version. http://www.bis.org/publ/bcbs128.htm (accessed 21 May 2010).

Ben Gal, I. (2007) Bayesian Networks, in *Encyclopaedia of Statistics in Quality and Reliability*, ed. F. Ruggeri, R.S. Kenett and F. Faltin, John Wiley & Sons, Ltd, Chichester.

Bonafede, E.C. and Giudici, P. (2007) Bayesian Networks for Enterprise Risk Assessment, *Physica A*, 382, 1, pp. 22–28.

Cornalba, C. and Giudici, P. (2004) Statistical Models for Operational Risk Management, *Physica A*, 338, pp. 166–172.

Cruz, M. (2002) *Modeling, Measuring and Hedging Operational Risk*, John Wiley & Sons, Ltd, Chichester.

Dalla Valle, L., Fantazzini, D. and Giudici, P. (2008) Copulae and Operational Risk, *International Journal of Risk Assessment and Management*, 9, 3, pp. 238–257.

Doebli, B., Leippold, M. and Vanini, P. (2003) From Operational Risk to Operational Excellence, http://ssrn.com/abstract=413720 (accessed 11 January 2010).

Fenton, N. and Neil, M. (2007) *Managing Risk in the Modern World: Applications of Bayesian Networks*, London Mathematical Society, London.

Figini, S., Kenett, R.S. and Salini, S. (2010) Integrating Operational and Financial Risk Assessments, *Quality and Reliability Engineering International*, http://services.bepress.com/unimi/statistics/art48 (accessed 6 March 2010).

Frachot, A., Georges, P. and Roncalli, T. (2001) Loss Distribution Approach for Operational Risk and Unexpected Operational Losses, http://ssrn.com/abstract=1032523 (accessed 21 May 2010).

Giudici, P. and Bilotta, A. (2004) Modelling Operational Losses: a Bayesian Approach, *Quality and Reliability Engineering International*, 20, pp. 407–417.

ICH (2006) The International Conference on Harmonization of Technical Requirements for Registration of Pharmaceuticals for Human Use, *Guidance for Industry: Q9 Quality Risk Management*, http://www.fda.gov/RegulatoryInformation /Guidances/ucm128050.htm (accessed 6 March 2009).

International Association of Financial Engineers (2010) http://www.iafe.org/html/cms_orc.php (accessed 8 March 2010).

ISO GUIDE 73 (2009) Risk management – Vocabulary.

Kaplan, R.S. and Norton, D.P. (1992) The Balanced Scorecard – Measures that Drive Performance, *Harvard Business Review*, 70, 1, pp. 71–79.

Kaplan, R.S. and Norton, D.P. (1993) Putting the Balanced Scorecard to Work, *Harvard Business Review*, 71, 5, pp. 134–142.

Kaplan, R.S. and Norton, D.P. (1996) *The Balanced Scorecard: Translating Strategy into Action*, Harvard Business School Press, Boston, MA.

Kenett, R.S. (2007) Cause and Effect Diagrams, in *Encyclopaedia of Statistics in Quality and Reliability*, ed. F. Ruggeri, R.S. Kenett and F. Faltin, John Wiley & Sons, Ltd, 2007.

Kenett, R.S. and Baker, E. (2010) *Process Improvement and CMMI for Systems and Software: Planning, Implementation, and Management*, Taylor & Francis Group, Auerbach Publications, Boca Raton, FL.

Kenett, R.S. and Zacks, S. (1998) *Modern Industrial Statistics: Design and Control of Quality and Reliability*, Duxbury Press, San Francisco.

KRIL (2010) The KRI Library, http://www.kriex.org/ (accessed 7 February 2010).

McKibben, D. and Furlonger, D. (2008) Magic Quadrant for Operational Risk Management Software for Financial Services, Gartner Industry, Research Note G00157289.

MUSING (2006) IST- FP6 27097, http://www.musing.eu (accessed 21 May 2010).

Neil, M., Fenton, N. and Tailor, M. (2005) Using Bayesian Networks to Model Expected and Unexpected Operational Losses, *Risk Analysis Journal*, 25, pp. 963–972.

Ograjenšek, I. and Kenett, R.S. (2008) Management Statistics, in *Statistical Practice in Business and Industry*, ed. S. Coleman *et al.*, John Wiley & Sons, Ltd, Chichester.

Shevchenko, P.V. (2009) Implementing Loss Distribution Approach for Operational Risk, *Applied Stochastic Models in Business and Industry*, DOI: 10.1002/asmb.811.

Solvency II (2009) http://www.solvency-2.com/ (accessed 21 May 2010).

Sunday Times (2010) Lehman's $50 Billion Conjuring Trick: A report into the American bank's collapse reveals financial chicanery and negligent management, March 14. Quoted from http://www.blacklistednews.com/news-7798-0-24-24–.html (accessed 21 May 2010).

Tapiero, C. (2004) *Risk and Financial Management: Mathematical and Computational Methods*, John Wiley & Sons, Inc., Hoboken, NJ.

The Foreign Exchange Committee (2004) Management of Risk Operational in Foreign Exchange, *Risk Analysis*, 25, 4, http://www.ny.frb.org/fxc/2004/fxc041105b.pdf (accessed 7 March 2010).

Part II

DATA FOR OPERATIONAL RISK MANAGEMENT AND ITS HANDLING

3

Ontology-based modelling and reasoning in operational risks

Christian Leibold, Hans-Ulrich Krieger and Marcus Spies

3.1 Introduction

In recent history, ontologies have become an accepted technology for enabling knowledge sharing. This is particularly the case in IT-intensive organizations, as they move away from the ad hoc understanding of data and information and reusability of data using simple documentation facilities. Ontologies allow, with the inclusion of semantics, automated reasoning to be implemented on observed unstructured data. This supports higher level decision processes, as it enables equivalent understanding of the modelled knowledge by humans and computers.

An ontology is defined as a set of definitions of concepts and their relationships. The basic relationship of identity implies the generality of one specific concept, a *superconcept*, over another, a *subconcept*, which allows inferring knowledge on the nature of the concepts (Bruijn, 2003). Further relationships describe properties that link classes corresponding to concepts. For example, an operational loss event can have the property that a specific IT system failure is associated with it.

The word 'ontology' is defined in *Webster's Revised Unabridged Dictionary* (1913) as: 'That department of the science of metaphysics which investigates and explains the nature and essential properties and relations of all beings, as such, or the principles and causes of being' (see www.dict.org). This points to philosophy

Operational Risk Management: A Practical Approach to Intelligent Data Analysis   Edited by Ron S. Kenett and Yossi Raanan   © 2011 John Wiley & Sons, Ltd

as the origin of the term. Nevertheless, it has been introduced to the area of artificial intelligence (AI) in applications such as natural language processing (NLP), information extraction (IE), knowledge engineering and representation, and has become a widespread notion (Studer *et al.*, 1998).

In the context of information and communications technology (ICT), an 'ontology' refers to a description of a part of the world in formal language that is used to facilitate knowledge sharing and reuse (Fensel, 2003). A widely accepted definition was introduced by Gruber (1993): 'An ontology is a formal explicit specification of a shared conceptualization.' Ontologies in this sense are often referred to as domain ontologies. The relationship between a conceptual model as provided in a domain ontology and generic philosophical conceptual models is studied in Spies and Roche (2007).

According to Fensel (2003), ontologies, through formal, real-world semantics and consensual terminologies, interweave human and machine understanding. Real-world semantics, although not fully captured by Gruber's definition, are a very important property of ontologies, which facilitates the sharing and reuse of ontologies among humans, as well as machines (computers).

Ontologies can be used as a backbone for the integration of expert knowledge and the formalization of project results, including advanced predictive analytics and intelligent access to third-party data, through the integration of semantic technologies.

Ontologies in the MUSING project (MUSING, 2006) define the structure of a repository of domain knowledge for services in financial risk management, operational risk management and internationalization. According to the basic software engineering practice of separation of concerns, domain knowledge should be kept separate from applications integrating various user functionalities. This enables the MUSING platform (i.e. the composition of ontology repository and specific applications) to adapt easily to changes in the underlying domain knowledge. As an example of this, consider the updated definition of a financial indicator. Instead of having to change all MUSING applications, only the domain knowledge needs to be updated in the repository and the applications will reflect the changes immediately by using indicator definitions queried from the repository. Ontologies are obviously tightly connected to the application domains. MUSING ontologies, for example, contain concepts related to financial risk management, operational risk management and internationalization services. In addition, for proper modelling of application domain concepts, there are established methodologies and frameworks that provide higher level structures used by the MUSING domain ontologies. Most importantly, in recent years, so-called upper ontologies have become increasingly important. Upper ontologies collect basic 'world knowledge' beyond specific application domains in a nearly standardized way. Using upper ontologies therefore:

- Enables domain ontologies to reuse a common pool of standard concepts and relations.

- Facilitates extensibility, combination and reuse of domain ontologies.

In the next section, through a case study we explain the modules of MUSING ontologies, starting from the conceptual model and the resulting layered structure of upper ontologies, and proceeding to particular models covering the specific area of operational risk management.

3.1.1 Modules

As mentioned in the introduction, a proper design of ontologies is needed so that they can be applied in the context of complex business applications. A further requirement for MUSING ontologies is that they have to be combined not only with business applications related to data mining, but also with computational linguistic applications for text annotation. This is used in the MUSING platform for the innovative integration of quantitative and qualitative data using textual descriptions from analysts of business entities and markets. The MUSING ontologies exploit domain-independent 'world knowledge' as contained in such textual data. In the following subsection, we examine this upper ontological part in the MUSING ontology modules.

3.1.2 Conceptual model

Figure 3.1 describes the logical architecture of the MUSING ontologies. The conceptual model incorporates the following distinct abstraction levels:

- Generic and axiomatic ontologies. Upper ontologies have been grouped here. The set of *ProtonAndSystem* ontologies comprises generic and axiomatic ontologies. The *ProtonExtension* ontology is the single point of access to this level. It provides the gateway to the temporal and upper level concepts that need to be generic for all MUSING ontologies – by importing the *Protime* ontology, which itself connects to time and adapted Proton ontologies. The temporal ontologies are *Protime, Allen*, 4*D* and *Time*; the Proton ontologies are *System, Top* and *Upper*. Further details will be given in Section 3.2.

- Domain independent ontologies. This layer covers MUSING-specific knowledge across domains such as *company* and *risk* ontology. Further details will be given in Section 3.3.

- Standard reference ontologies, to make standards available in MUSING applications. The emerging standard reference level consists of *Industrial-Sector* representing NACE code, *BACH* and *XBRL-GAAP* suitable to work for example with the German HGB. See Section 3.4.1.

All of these upper level ontologies are used across domains in specific applications:

- Pilot-specific ontologies used in applications. MUSING services implement specific ontologies at a pilot-specific level. An example is

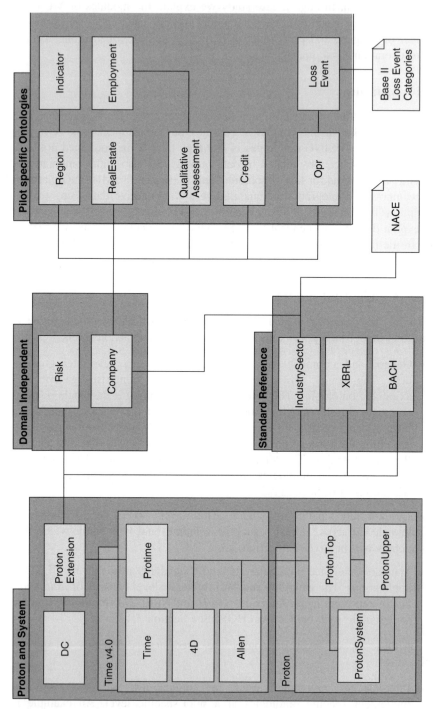

Figure 3.1   MUSING ontologies conceptual model.

QualitativeAnalysis. This ontology extends the domain covered by the *Company* ontology and reuses concepts of internationalization ontologies (e.g. *Employment, RealEstate*). This set of ontologies is completed by five ontologies, namely *Indicator, Region, Credit, OperationalRisk* and *LossEvent*.

The task of ontology coding transfers the captured conceptualizations into a formal ontology language. The ontology engineering methodology applies a middle-out ontology construction approach that identifies and specifies the most important concepts and relationships first. Then, the composed ontology becomes a basis to obtain the remainder of the hierarchy by generalizations and specializations. The approach used is an iterated ontology capture and ontology coding process that allows for frequent feedback from domain experts. The iterative processes terminate when all formal competency questions are encoded properly into the target ontology.

As a compromise between expressivity and reasonability, OWL DLP (Grosof *et al.*, 2003) is used as the ontology language within the MUSING project. MUSING ontologies are limited to the use of OWL DL syntax, where possible even OWL Lite, in order to maximize performance for reasoning tasks while trading off the required expressivity of the formal language used (see www.w3.org/TR/owl-semantics).

Figure 3.2 shows the complete import diagram of the MUSING ontologies. It reflects the matured understanding of technical coherences in the project after a reorganization of ontology levels. The import structure is chosen in order to allow the reuse of ontology schema from previous releases of the MUSING ontologies in the widest possible scale. This is especially relevant to allow the use of the ontologies across several versions and benefits from the improved conceptual model and refined structure of the knowledge.

The illustrated import hierarchy of the MUSING ontologies shows the key function of the *Extension* ontology (pext, a Proton extension) as the gateway to axiomatic and generic ontologies as well as the clear separation of domain-independent and domain-dependent ontologies. All implemented standard ontologies are available to the *Company* ontology, a fact which facilitates the standard-compliant processing of information from the underlying information structures.

Figure 3.3 provides an analysis of an intermediate state of implementation in terms of cardinal numbers. The MUSING version 1.0 ontologies consist of 3404 concepts, 3033 properties and 1461 instances. A large number of these are inherited from the inclusion of the standard reference ontologies – XBRL-GAAP has 2700 concepts alone. Of course, this number changes with updates of the standards, see www.webcpa.com/news/FAF-Will-Now-Maintain-XBRL-GAAP-Taxonomy-53227-1.html (accessed 21 May 2010).

Upper ontologies provide world knowledge in 252 concepts and 138 properties. Time is represented in 111 concepts and 77 properties. Domain-independent ontologies shape out the business intelligence frame with 233 concepts and 65

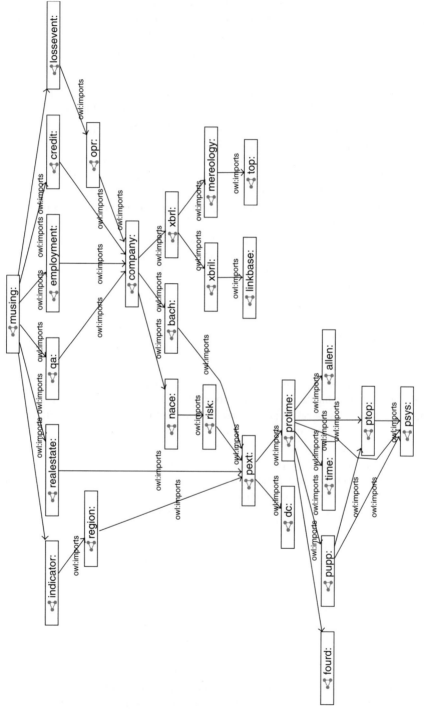

Figure 3.2   Complete import graph of MUSING ontologies.

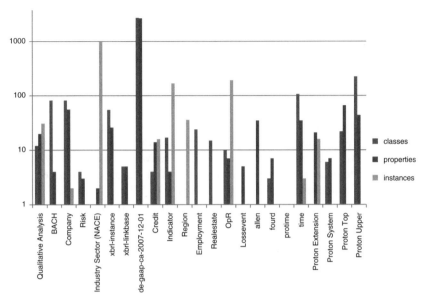

Figure 3.3   Distribution of concepts, properties and instances over the MUSING ontology schema set.

properties. Operational risk-specific specializations, and subconcepts, sum to 216 of the above-mentioned elements and are described further in Section 3.5. The remaining 452 concepts and properties are used to specify particular knowledge in the other application areas.

In the following sections we describe the MUSING ontologies of version 1.0 in detail, with special attention to their relevance for operational risk management.

3.2   Generic and axiomatic ontologies

The generic and axiomatic ontologies include the set of adapted Proton ontologies and the temporal ontologies, concepts which are common to all MUSING ontologies. This level includes Proton *System, Top* and *Upper* (Terziev *et al.*, 2005), *DublinCore* (http://dublincore.org) and the temporal ontologies (*Protime, Allen*, 4D and *Time*).

In Section 3.2.1, we describe the changes to the *ProtonExtension* ontology and in Section 3.2.2, the temporal ontologies, internal version T5.0. This layer uses *DublinCore* ontology and the three Proton ontology modules *System, Top* and *Upper*, which have not been modified from their original publication.

3.2.1   Proton extension

MUSING ontologies have their axiomatic foundation in the *ProtonExtension* ontology. This foundation is used in order to tailor Proton ontologies to the needs of the business intelligence (BI) domain and the specific requirements set by the different tasks involving the ontology as knowledge repository (e.g. regarding data representations).

In order to facilitate the integration of different schemata and data items, the properties listed in Table 3.1 have been added. Properties for *pupp:PostalAddress* are included in order to allow smooth integration and reuse of information extracted in the several extraction tasks. This representation overwrites the PostalAdress as a geographical location, which was of no use as global positions of companies are seldom available in documents and not commonly used as a reference in BI. The specification of *pext:hasFirstName* and *pext:hasLastName* allows the reference to a person in case the full name is not given in the documents to which the extraction task is applied. It is anyway useful to have both the full name and the 'structured' parts of the naming of a person.

Table 3.1   Additions to the *ProtonExtension* ontology.

New property for *ptop:Agent* (e.g. for Company)	
pext:hasFullName	\<string\>
New properties for *ptop:Person*	
pext:hasFirstName	\<string\>
pext:hasLastName	\<string\>
pext:hasNationality	\<multiple *pupp:Country*\>
New properties for *pupp:PostalAddress*	
pext:hasCity	\<string\>
pext:hasCountry	\<string\>
pext:hasPostCodeMajorClient	\<int\>
pext:hasProvince	\<string\>
pext:hasStreetName	\<string\>
pext:hasStreetNumber	\<string\>
pext:hasZipCode	\<int\>

Upcoming work includes lifting the concept *Measurements* to the axiomatic level. For this *ProtonExtension* is a candidate for the new target ontology. The measurements apply to risk and regional indicators. Originally used as application-specific variables, these indicators are exploited throughout the MUSING domains as qualified knowledge and validated through the pilot application results.

3.2.2   Temporal ontologies

The temporal ontologies are used to model the temporal dimension within the scope of the MUSING ontology modelling efforts. Following the conceptual model, only the *Extension* ontology imports the temporal ontologies. Time issues are also discussed thoroughly in Krieger *et al.* (2008).

The 4D ontology defines the key concepts needed to model time issues, which are *fourd:Perdurant, fourd:TimeSlice* and *fourd:Time*. Furthermore it defines related properties such as *fourd:hasTimeSlice* and *fourd:hasTime*.

The *Time* ontology defines the class *time:TemporalEntity* and its subclasses. The ontology is compatible with OWL-Time described in Hobbs and Pan (2004) by reusing the class names of OWL-Time for *time:TemporalEntity, time:Instant* and *time:Interval* as well as the properties *time:begins* and *time:ends*. This approach saved the trouble of creating a set of equivalence axioms.

Figure 3.4 illustrates the incorporation of the time dimension into the set of MUSING ontologies. All classes of the VS-independent and VS-dependent ontologies such as *ViceChairman* and *Chairman* are subclasses of the Proton *System* module class *psys:Entity* which is equivalent to the class *fourd:TimeSlice*. Thus, every entity has a property *fourd:hasTime* specifying an instant or an interval in which certain properties of the entity hold. In the example, the instance 'John' of the class *fourd:Perdurant* has two time slices. From 2003 to 2005 John was vice chairman and from 2005 till now he is chairman.

It is often required to assign multiple time slices of the same type to an instance of the class *fourd:Perdurant*. For example, John could already have

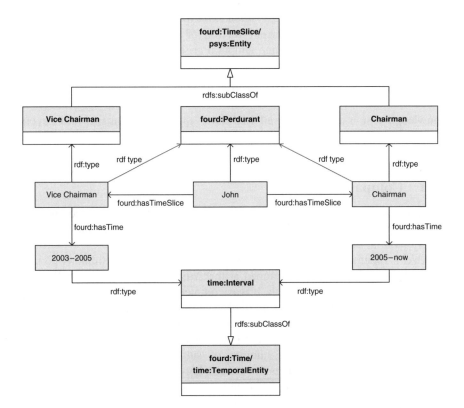

Figure 3.4   Inclusion of the time dimension into the MUSING ontologies.

been chairman in the period from 2003 to 2005 but with a different salary, say. Also note that the time slices of one *Perdurant* may overlap.

The group of temporal ontologies has been extended by the *Protime* and *Allen* ontologies.

The *Protime* ontology provides a **single point of access to the upper level ontologies and temporal axioms**. These include the distinction between duration time and calendar time. Whereas the latter allows the mapping of calendar items to ontology concepts, duration time is represented as additive durations. In this context some concepts have been renamed (e.g. *Day* → *CalendarDay*) in order to facilitate readability and improve distinction.

The *time:Duration* concept uses the functional properties *noOfYears, noOf-Months, noOfDays, noOfHours, noOfMinutes* and *noOfSeconds* to compose the respective duration.

In order to allow and facilitate reasoning on temporal information we provide an additional temporal dimension (partial ordering) on underspecified time intervals. This is based on Allen's interval logic (Allen, 1983). Allen defines 13 temporal topological relations on intervals; we define them on time slices. These relations are *allen:equals, allen:finishedBy, allen:finishes, allen:after, allen:before, allen:contains, allen:during, allen:meets, allen:metBy, allen:overlaps, allen:overlappedBy, allen:startedBy* and *allen:starts*.

As an extra feature, the current version of time ontology adds definitive durations (temporal facts) to define temporal arithmetic.

3.3   Domain-independent ontologies

Knowledge that is common to all MUSING application domains and beyond is combined and represented in this ontological layer. The core ontology is the company ontology, described in the following subsection.

3.3.1   Company ontology

A comprehensive survey of generic enterprise models is presented in Belle (2002). In the following, we briefly describe the main enterprise models which are available in ontological form.

AIAI's Enterprise Ontology is a collection of terms and definitions pertaining to business enterprises, developed at the Artificial Intelligence Applications Institute, University of Edinburgh. It was completed in 1996 in natural language format and ported to Ontolingua in 1998. The ontology is discussed thoroughly in Uschold *et al.* (1998). See www.aiai.ed.ac.uk/project/enterprise for more.

TOVE (The Toronto Virtual Enterprise Ontology) is a partially completed ontology which consists of a number of sub-ontologies, with the aim of creating a formal and computable terminology and semantics of activities and resources in the realm of the enterprise. It is an ongoing project at the Enterprise Integration Laboratory, University of Toronto (Fox, 1992). See www.eil.utoronto.ca/enterprise-modelling for more.

The CYC ontology is a project that attempts to formalize common sense. Built on a knowledge base core with millions of assertions, it attempts to capture a large portion of consensus knowledge about the world. More than 600 concepts of the ontology are directly related to the enterprise domain. The CYC ontology is available from www.cyc.com/.

MUSING applies the Enterprise Ontology developed by the Artificial Intelligence Applications Institute at the University of Edinburgh. This choice was made because of the wide scope of the Enterprise Ontology, which is also well documented and easily accessible.

The *Company* ontology describes a fragment of the economy from the perspective of a single company. The *Company* ontology of the MUSING ontologies (since version 0.6) relies on the Enterprise Ontology version 1.1, which represents a collection of terms and definitions relevant to business enterprises. The Enterprise Ontology was developed in the Enterprise Project by the Artificial Intelligence Applications Institute at the University of Edinburgh with IBM, Lloyd's Register, Logica UK Limited and Unilever as partners. The project was supported by the UK's Department of Trade and Industry under the Intelligent Systems Integration Programme. For further information about the Enterprise Ontology refer to Uschold *et al.* (1998).

For various reasons not all the concepts and relations of the Enterprise Ontology are available in the MUSING *Company* ontology. As an example with respect to the representation of temporal concepts, it was necessary to derogate from the Enterprise Ontology.

Conceptually, the MUSING *Company* ontology as well as the Enterprise Ontology are divided into four main parts, namely Activities and Processes, Organization, Strategy and Marketing.

The core class of the **Activities and Processes** part is *company:Activity*. An activity captures the notion of anything that involves doing. The concept is closely linked with the idea of the doer, which may be a person, organizational unit or machine modelled as *company:PotentialActor*. The property *company:haveCapability* denotes the ability (or skill if the doer is a person) of a potential actor to be the doer of an activity. Actors may also have other roles with respect to an activity such as activity owner.

Figure 3.5 shows the relations of the class *company:Activity*. According to the ontology, an activity can have an activity status, effects, outputs and preconditions, and it can use resources. Large and complex activities that take a long time may be composed of a set of sub-activities. An activity is the execution of a specific activity specification. Activities can also be owned entities within the scope of an ownership situation.

Activity specifications which are instances of the class *company:ActivitySpec* specify one or more possible events at some level of detail. If an activity specification has an intended purpose, it is called a plan. For plans the class *company:Plan* is used. A plan which is executed repeatedly is represented as a process specification in terms of the class *company:ProcessSpec*.

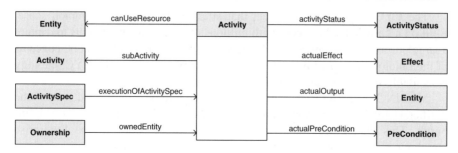

Figure 3.5   Relations of the company:Activity *class.*

The property *company:holdAuthority* is used to denote that an actor has the right to perform the activities specified in an activity specification.

Central to the **Organization** part are the concepts of legal entity and organizational unit (OU). They differ in that a legal entity is recognized as having rights and responsibilities in the world at large and by legal jurisdictions in particular, whereas OUs only have full recognition within an organization. Corporations as well as persons are considered legal entities.

Figure 3.6 illustrates the relations of classes to the class *company:LegalEntity*. A legal entity can be a customer and vendor in a potential or actual sale situation. Legal entities can own other legal entities. With respect to a shareholding situation, a legal entity can be a shareholder or the respective corporation that is held. By means of an employment contract, a legal entity can be connected as an employer with its employees. Employees are persons who can also be partners in a partnership situation. Persons who are employees typically work for an OU. OUs can be large and complex, even transcending legal entities, and may be made up from smaller ones. It is also common for an OU to manage other OUs.

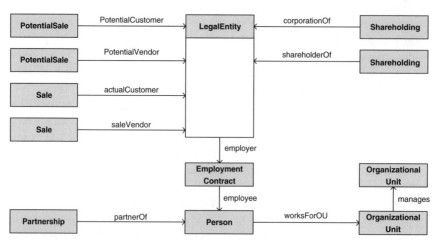

Figure 3.6   Relations of the classes company:LegalEntity *and* company: OrganizationalUnit.

The smallest OU may correspond to a single person. In fact, a particular person can also be seen as corresponding to more than one OU.

A further concept defined in this part of the *Company* ontology is the concept of machine. A machine is neither a human nor a legal entity, but can play certain roles otherwise played by a person or OU.

The ownership of rights and responsibilities can, from a legal point of view only, lie with a legal entity. However, within organizations, rights and responsibilities are allocated to OUs. For this reason, legal ownership and non-legal ownership were defined to enable the distinction where needed. OUs are typically responsible for certain activities.

The developers of the Enterprise Ontology intended to represent the management structure within an organization by management links. The property *company:manage* is used to assign purposes to OUs. The management structure is defined as a pattern of management links between OUs. This includes multiple management links into any OU with constraints on the purposes assigned through each link.

The central concept of the **Strategy** part is purpose. Figure 3.7 shows the different uses of the class *company:Purpose*. A purpose can be either intended by an activity specification or held by an OU. In the prior case, the purpose captures the idea of something which a plan can help to achieve. In the latter case, the purpose defines what an OU is responsible for. The description of a purpose can be of any kind, whether strategic or operational. One statement of purpose can relate to something which can also be seen to help to achieve some grander purpose. This means that a purpose can be composed or decomposed.

Figure 3.7   Relations of the company:Puropose *class.*

Strategy is defined as a plan to achieve a high-level purpose. The respective concept is represented by the class *company:Strategy*. Based on the concept of a plan, strategic planning can be represented by the terms decision, assumption, risk and various types of factor.

The central concept of the **Marketing** part is sale. Figure 3.8 describes the relations of the *company:Sale* class. A sale situation typically has legal entities of customer and vendor. The price and the product can be a good, a service or money. This means that a sale is an agreement between two legal entities for the

Figure 3.8   Relations of the company:Sale *class.*

exchange of a product for a price. Usually the product is a good or a service and the price is monetary, but other possibilities are included. A sale could have been agreed in the past and a future potential sale can be envisaged, whether or nor the actual product can be identified or even exists. A potential sale is represented by the class *company:PotentialSale*.

The market is represented by all sales and potential sales within the scope of interest and may include sales by competitors. A market can be decomposed into market segments. For the analysis of a market it may be useful to involve the understanding of product features, needs of customers and images of brands, products or vendors. Promotions are activities whose purposes relate to the image in a market.

The *Risk* ontology defines the basic concept of risk for the MUSING project. Core properties are *Severity, Frequency* and *Distribution*. Furthermore, the ontology allows us to model correlated *Control*(flows), *MitigationStrategy* and *RiskProfiles* for each observed risk. These concepts can be used as the backbone for complex and highly specific risk definitions, being generic enough to allow freedom of design for the particular case while ensuring coverage of the vital topic and correlating them with each other. The example for the case of specialization in operational risk is given in Section 3.5.

3.4   Standard reference ontologies

This ontological layer contains concepts that are relevant for contextualizing the MUSING knowledge and applications to standards. An ontology in this layer includes references to or representations of standards that are themselves either mandatory in order to comply with regulatory requirements (e.g. XBRL), or agreed best practices (e.g. BACH, NACE).

3.4.1 XBRL

XBRL (eXtensible Business Reporting Language) defines schemata to be used in business reporting, especially in reporting to stock markets or regulatory offices. Thus the underlying machine-readable format allows exploitation of the available data for several other reporting and business cases, within or outside of a company.

The namespaces XBRL 'Instance' and XBRL 'Linkbase' are designed to represent the structure of the XBRL schema files (instance and linkbase) suitably for the MUSING ontologies.

MUSING uses a model-based transformation approach to translate from schema to ontology format that allows the integration and access from other concepts of the MUSING ontological family. For details, see Spies (2010).

The structure of the instance documents of XBRL is defined in the schema document for the namespace www.xbrl.org/2003/instance which is available online at www.xbrl.org/2003/xbrl-instance-2003-12-31.xsd.

The resulting XBRL ontology contains all relevant knowledge to access XBRL structural information from the linkbase and specific components reported by a company in an instance document. It explains the relation between tuples and their components. This represents the major impact of the current XBRL ontology implementation.

The W3C interest group working towards the representation of XBRL in RDF (Resource Description Framework) and the availability of transparent finance data on the Web.

3.4.2 BACH

The *BACH* ontology relies on the Bank for the Accounts of Companies Harmonized (BACH) database scheme. It is an attempt to allow for the interoperability of accounting data on a European level. BACH contains concepts that are related to accounting and controlling tasks – therefore this ontology is central to the identification of financial risks.

3.4.3 NACE

The *IndustrySector* ontology is based on NACE, a European industry standard classification system to describe the activity areas of companies in a specific industry sector. The various elements are modelled as instances of the Proton *Upper* ontology class *pupp:IndustrySector*. The OWL code of a sample element is shown in Listing 3.1.

Listing 3.1   Sample of an instance of *pupp:IndustrySector*.

```
<pupp:IndustrySector rdf:ID="nace_11.03">
 <pupp:subSectorOf rdf:resource="#nace_11.0"/>
 <hasLevel
  rdf:datatype=".../XMLSchema#int">4</hasLevel>
 <hasCode rdf:datatype=".../XMLSchema#string">11.03
 </hasCode>
 <rdfs:label xml:lang="en">Manufacture of cider and other
  fruit wines</rdfs:label>
 <rdfs:label xml:lang="de">Herstellung von Apfelwein und
  anderen Fruchtweinen</rdfs:label>
 <rdfs:label xml:lang="it">Produzione di sidro e di altri
  vini a base di frutta</rdfs:label>
</pupp:IndustrySector>
```

The *rdf:ID* of each *pupp:IndustrySector* instance is the concatenation of 'nace_' and the identifier of the NACE element. The *pupp:subSectorOf* property is used to represent the hierarchy of the instances. The datatype properties *nace:hasLevel* and *nace:hasCode* indicate the depth in the hierarchy and the identifier of the NACE element, respectively. The NACE elements are classified on four levels. Finally, *rdfs:label* properties are used to describe the sector of the NACE element in English, German and Italian.

The *IndustrySector* ontology contains instances of 996 elements of the NACE code.

3.5   Operational risk management

3.5.1   IT operational risks

The purpose of ontologies in the domain of IT operational risks is to build a repository of risk models according to statistical research paradigms based on relevant legislation and standards.

The legislation and standards concerned with IT operational risks are mostly related to the Basel II Agreement (Basel Committee on Banking Supervision, 2006) that contains loss event classifications for several risk categories and prescribes appropriate measures to be taken by companies. However, Basel II is not specifically concerned with IT operational risk. IT operational risk models should be expressed in terms of:

- Areas (relevant sectors of business activity).

- Causes (roughly classified into people related/external events/processes/ systems failures).

- Types of loss incurred (internal/external), further qualified by actual data (internal source or provider database) and/or expert opinion.

The IT operational risk data consists mainly of frequency and severity of loss events, which can be expressed in exploratory studies as separate ratings or, in models based on statistical inference, as a probability distribution for a particular area and given causes.

In this section, we discuss the ontology approaches to loss event types and area modelling.

The **loss event type** classification in Annex 1 of a document provided by the Basel Committee on Banking Supervision titled 'QIS 2 – Operational Risk Loss Data – 4 May 2001' (Basel Committee on Banking Supervision, 2001) has influence on the IT operational risk ontology at two distinct points.

First, the *risk:ITOperationalRisk* class has subclasses corresponding to the first two levels of the classification. Second, the *itopr:LossEvent* class is subclassed by classes representing the loss event type classification. This construction makes it possible to encode the type or risks as well as loss events in the class hierarchy.

Currently, the other two attributes of risks and loss events, namely cause and area, are specified with the help of the two object properties *hasCause* and *hasArea*. Since areas represent specific industries or economic activities, the range of the respective property consists of all instances of *pupp:IndustrySector*.

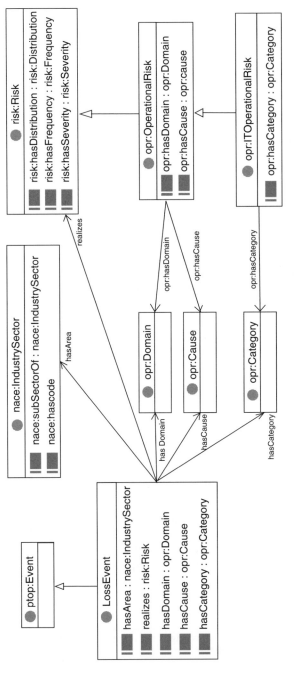

Figure 3.9   Central concepts of the (IT) operational risk domain ontologies.

For the time being, 996 items of the NACE classification have been instantiated. According to Basel II, operational risk is defined as the risk of loss resulting from inadequate or failed internal processes, people and systems, or from external events. Hence, the *itopr:Cause* class was instantiated accordingly.

In comparison with the loss event instances, the risk instances inherit a *has-Distribution* object property from the *risk:ITOperationalRisk* class. As usual in statistics, instances of the *itopr:Distribution* class have attributes indicating their type and parameters.

Turning to the **business areas**, MUSING decided to use the NACE code. As described above, the NACE code is a classification of economic activities in the European Community. BACH reports data according to the NACE rev1.1 classification at the one-letter, two-letter and two-digit levels. That is why this particular subset of NACE has been ontologized so far. During the implementation, the grounding into PROTON could again be exploited: this framework already includes a class called *IndustrySector* which defines the properties *has-Code* and *subSectorOf*. Based on this, the actual NACE sectors have been created as instances of *IndustrySector*. As well as an identifier and a label, each of these 996 instances has a string value representing the actual code and an object property pointing to – if applicable – the respective supersector. Also, the link between companies and sectors is already given within PROTON: each PROTON company derives the property *activeInSector* – which has *Industry-Sector* defined as its range – from the superclass *CommercialOrganization*.

The business areas ontology approach is also relevant to internationalization services. The structure of the related ontologies is illustrated in Figure 3.9.

The core concepts relevant to the representation of risks, specific to a company operating in a specific business area, are inherited through the ontologies *Company* and *IndustrySector*.

3.6   Summary

This chapter presents the general application of ontologies as foundation for the representation of knowledge. We used, as a case study, the specific knowledge models developed in the domain of operational risks in MUSING. The conceptual separation of ontologies inherently follows the logical division of the scope and context of the specific knowledge in the project. The resulting MUSING ontology set is a blend of existing, validated ontologies, knowledge formalized in standards and specific requirements from the project applications.

References

Allen, J. F. (1983) Maintaining knowledge about temporal intervals. *Communications of the ACM*, 26, 11, pp. 832–846.

Basel Committee on Banking Supervision (2001) *QIS 2 – Operational Risk Loss Data*, http://www.bis.org/bcbs/qisoprisknote.pdf (accessed 21 May 2010).

Basel Committee on Banking Supervision (2006) *Basel II: International Convergence of Capital Measurement and Capital Standards: A Revised Framework – Comprehensive Version*, http://www.bis.org/publ/bcbs128.htm (accessed 11 January 2010).

Belle, J. (2002) A survey of generic enterprise models. Paper presented at *SACLA, 2002*.

Bruijn, J. (2003) Using ontologies – enabling knowledge sharing and reuse on the semantic web. Technical Report DERI-2003-10-29, DERI.

Fensel, D. (2003) *Ontologies: A Silver Bullet for Knowledge Management and Electronic Commerce*, Springer-Verlag, Berlin.

Fox, M. S. (1992) The TOVE Project: Towards a common-sense model of the enterprise. Technical Report, Enterprise Integration Laboratory.

Grosof, B. N., Horrocks, I., Volz, R. and Decker, S. (2003) Description logic programs: combining logic programs with description logic. Paper presented at *the 12th International Conference on the World Wide Web*, Budapest, Hungary.

Gruber, T. R. (1993) A translation approach to portable ontology specifications. *Knowledge Acquisition*, 5, pp. 199–220.

Hobbs, J. R. and Pan, F. (2004) An ontology of time for the Semantic Web. *ACM Transactions on Asian Language Processing* (TALIP), 3, 1, pp. 66–85.

Krieger, H.-U., Kiefer, B. and Declerck, T. (2008) A Framework for temporal representation and reasoning in business intelligence applications. *AAAI Spring Symposium*, pp. 59–70).

MUSING (2006) IST- FP6 27097, http://www.musing.eu (accessed 21 May 2010).

Spies, M. (2010) An ontology modeling perspective on business reporting. *Information Systems*, 35, 1, pp. 404–416.

Spies, M. and Roche, C. (2007) Aristotelian ontologies and OWL modeling. In P. Rittgen (Ed.), *Handbook of Ontologies in Business Interaction*, Idea Group, Hershey, PA, pp. 21–29.

Studer, R., Benjamins, V. R. and Fensel, D. (1998) Knowledge engineering: principles and methods. *Data and Knowledge Engineering* (DKE), 25, 1/2, pp. 161–197.

Terziev, I., Kiryakov, A. and Manov, D. (2005) Base upper-level ontology (BULO) (Project Deliverable), SEKT.

Uschold, M., King, M., Moralee, S. and Zorgios, Y. (1998) The Enterprise Ontology. *Knowledge Engineering Review*, 13, pp. 31–89.

4

Semantic analysis of textual input

Horacio Saggion, Thierry Declerck and Kalina Bontcheva

4.1 Introduction

Business intelligence (BI) can be defined as the process of finding, gathering, aggregating and analysing information for decision making (Chung *et al.*, 2003). However, the analytical techniques frequently applied in business intelligence (BI) have been largely developed for dealing with numerical data, so, unsurprisingly, the industry has started to struggle with making use of distributed and textual unstructured information.

Text processing and natural language processing (NLP) techniques can be used to transform unstructured sources into structured representations suitable for analysis. Information extraction (IE) is a key NLP technology which automatically extracts specific types of information from text to create records in a database or to populate knowledge bases. One typical scenario for information extraction in the business domain is the case of insurance companies tracking information about ships sinking around the globe. Without an IE system, company analysts would have to read hundreds of textual reports and manually dig out that information. Another typical IE scenario is the extraction of information about international and domestic joint ventures or other types of firms' agreements from unstructured documents. This kind of information can help identify

Operational Risk Management: A Practical Approach to Intelligent Data Analysis   Edited by Ron S. Kenett
and Yossi Raanan   © 2011 John Wiley & Sons, Ltd

not only information about who is doing business with whom and where, but also market trends, such as what world regions or markets are being targeted by which companies and in which industrial or service sector.

One additional problem with business information is that, even in cases where the information is structured (e.g. balance sheets), it may not be represented in a way machines can understand – and this is particularly true with legacy systems and documentation. One response to this problem has been the development of the emerging standard XBRL (eXtensible Business Reporting Language) for reporting financial information.

This chapter describes a number of tools for text analysis in business intelligence with the General Architecture for Text Engineering (GATE). For more on GATE see Cunningham *et al.* (2002).

4.2   Information extraction

IE is the mapping of natural language texts (e.g. news articles, web pages, emails) into predefined structured representations or *templates*. These templates represent the key information a user has specified as important for extraction and they are therefore dependent on the particular task or scenario.

IE is a complex task carried out by human analysts on a daily basis. Because it is very time consuming and labour intensive, there has been much research over the last 20 years to automate the process. Research was stimulated by a series of competitions from 1987 to 1997 known as MUCs (Message Understanding Conferences). As systems began to achieve very high results over closed domains (such as news texts about company takeovers), research turned towards various new directions. First, a more semantically based approach was adopted, whereby IE became more a task of *content extraction*, as witnessed by programs such as ACE (see www.itl.nist.gov/iaui/894.01/tests/ace/) which dealt with the semantic analysis of text rather than the linguistic analysis imposed by the MUC competitions. Second, the need arose for systems which can be quickly and easily tailored to new domains, languages and applications (Maynard *et al.*, 2003a). The TIDES Surprise Language Exercise is an excellent example of this (see www.darpa.mil/iao/TIDES.htm). Here participants were given language analysis tasks such as machine translation and information extraction on a selected language with no advance warning of what this language might be, and a deadline of a month to develop tools and applications from scratch for this language. See for example Maynard *et al.* (2003b) for details of an IE system developed during this task. Third, the emergence of the Semantic Web entailed the need for combining IE with information stored in ontologies, in order to gain a more detailed semantic representation of the extracted information (see Bontcheva *et al.*, 2004).

IE is not the same as information retrieval. Information retrieval pulls documents from large text collections (usually the Web) in response to specific keywords or queries. Usually there is little linguistic analysis of these keywords or queries, and simple string matching is performed on the query. IE, on the other hand, pulls facts and structured information from the content of large text collections without the user specifying a particular query in advance. The user then analyses the facts. In this case, detailed linguistic analysis is generally performed on the text in order to extract the most appropriate parts.

With traditional search engines, getting the facts can be hard and slow. Imagine you wanted to know all the places the Queen of England had visited in the last year, or which countries on the east coast of the United States have had cases of West Nile Virus. Finding this information via a search engine such as Google would be virtually impossible, as there is no easy way to select the appropriate search terms, and it is likely that this information would be spread over a wide variety of documents and would therefore involve a number of consecutive searches. IE techniques, however, could easily find the answers to these questions by identifying keywords such as 'the Queen', 'visit' and 'last year' in the same sentence, or by flagging all the names of diseases, locations, etc., enabling a much simpler search. In the finance field, IE techniques can quickly flag up important terms such as names of countries, cities, organisations, time periods, addresses, share prices, and so on.

There are two main approaches to the development of IE systems: (1) hand-crafted systems which rely on language engineers to design lexicons and rules for extraction; and (2) machine learning systems which can be trained to perform one or more of the IE tasks such as named entity recognition and co-reference resolution. Learning systems are given either an annotated corpus for training or a corpus of relevant and irrelevant documents together with only a few annotated examples of the extraction task. In this case, some non-supervised techniques such as clustering can also be applied. Rule-based systems can be based on gazetteer lists and cascades of finite state transducers. Gazetteer lists are lists of keywords which can be used to identify known names (e.g. New York) or give contextual information for recognition of complex names (e.g. Corporation is a common postfix for a company name). Transducers implement pattern matching algorithms over linguistic annotations produced by various linguistic processors (Cunningham *et al.*, 2002; Appelt *et al.*, 1993).

Symbolic learning techniques which learn rules or dictionaries for extraction have been applied in IE. The AutoSlog system (Riloff, 1993) (and later the AutoSlog-TS system) automatically constructs a dictionary of extraction patterns using an instantiation mechanism based on a number of syntactic templates manually specified, a corpus syntactically parsed and a set of target noun phrases to extract. LP2 identifies start and end semantic tags (the beginning and end tokens of a concept in text) using a supervised approach (Ciravegna, 2001). LP2 learns three types of rules: tagging rules, contextual rules and correction rules. The key to the process is in the separation of the annotations into start and end annotations (e.g. beginning of a date annotation and end of a date annotation) and in the exploitation of the interactions between rules which identify start and end annotations (e.g. some rules may include information about start/end tags). ExDISCO learns extraction patterns (which then have to be associated with templates slots) from a set of documents (Yangarber *et al.*, 2000). Statistical machine learning approaches to information extraction include the use of hidden Markov models (HMMs), support vector machines (SVMs), and conditional random fields (CRFs).

With HMMs the IE task is cast as a tagging problem (Leek, 1997). Given a sequence of input words, the system has to produce a sequence of tags where the words are observations and the tags are hidden states in the HMM. SVMs are very competitive supervised models for IE casting the IE task as a binary classification problem, each label giving rise to a binary classification problem (Isozaki and Kazawa, 2002). SVMs try to find a hyperplane in the vector space of instances that maximally separates positive from negative instances. Finding the hyperplane corresponds to an optimisation problem. CRFs are state-of-the-art techniques for IE and tend to do better than other classification methods (Lafferty *et al.*, 2003).

4.2.1   Named entity recognition

Named entity recognition forms the cornerstone of almost any IE application. It consists of the identification of important entities such as proper names in text and their classification into predefined categories of interest. Traditional types of named entities, as used in the MUC competitions mentioned previously, are person, location, organisation, monetary items, dates and times. Other kinds of entities may also be recognised according to the domain, such as addresses, phone numbers, URLs, names of journals, ships, exchange rates and so on.

Named entity recognition is important because it provides a foundation from which to build more complex IE systems. For example, once the entities have been found, we can look at identifying relations between them, which may help with co-reference, event tracking, scenario building, and so on.

Approaches to named entity recognition generally fall into one of two types: the knowledge engineering approach or the machine learning approach. Knowledge engineering approaches are rule based, and must be developed by experienced language engineers or linguists. These rely on hand-coded rules which identify patterns in the text and make use of human intuition. They can be quite time consuming to develop for the user. Furthermore, it can be difficult to find a suitably qualified person who is also knowledgeable about the domain in question. For example, most linguists would have difficulty writing rules to find names of genes and proteins in the biology domain, unless they had help from domain experts. Machine learning techniques, on the other hand, use statistical methods to identify named entities, and do not require human intervention. Providing that a large corpus of training material is provided, these systems can work very well and are inexpensive to develop and run. But herein lies the catch: the creation of such training data is difficult and time consuming to produce, as it requires a domain expert first to annotate manually large volumes of data for the system to train on. In general, the more the available training data, the better the system.

4.3   The general architecture for text engineering

The General Architecture for Text Engineering (GATE) is a freely available open source architecture for language processing implemented in Java (Cunningham *et al.*, 2002). GATE consists of a framework for creating, adapting and deploying human language technology (HLT) components and is provided with a Java library and Application Programming Interface (API) for system development.

GATE provides three types of resources: language resources (LRs) which collectively refer to data; processing resources (PRs) which are used to refer to algorithms; and visualisation resources (VRs) which represent visualisation and editing components provided in a graphical user interface (GUI).

GATE can be used to process documents in different formats including plain text, HTML, XML, RTF, SGML and PDF. When a document is loaded or opened in GATE, a document structure analyser is called upon which is in charge of creating a GATE document. This is an LR which contains the text of the original document and one or more sets of annotations, one of which will contain the document markup annotations (e.g. HTML tags), if such exist. Annotations are generally updated by PRs during text analysis, but they can also be created manually in the GATE GUI. Each annotation belongs to an annotation set and has a type, a pair of offsets (denoting the span of text annotated) and a set of features and values that are used to encode further information. Features (or attribute names) are strings, and values can be any Java object. Features and values can be specified in an annotation schema which facilitates validation and input during manual annotation. Annotations and feature values created during document processing can be used for example to populate database records or as the basis for ontology population.

GATE comes with a default application for information extraction, called ANNIE, which consists of a set of core language processing components such as tokeniser, gazetteer lookup, part-of-speech tagger, and grammars for semantic analysis and annotation. ANNIE is rule based, which means that unlike machine-learning-based approaches, it does not require large amounts of training data. On the other hand, it requires a developer to create rules manually. Machine learning IE is also available in GATE through a plug-in which implements SVM classification for problems of text classification and concept recognition.

One key component in GATE is the Java Annotation Pattern Engine (JAPE), which is a pattern matching engine implemented in Java (Cunningham *et al.*, 2000). JAPE uses a compiler that translates grammar rules into Java objects that target the GATE API, and a library of regular expressions. JAPE can be used to develop cascades of finite state transducers.

Figure 4.1 shows a screenshot of a text processed by GATE's default IE system ANNIE. Each entity type (e.g. Location, Organisation, Person, etc.) is colour coded, and mentions of each entity in the text are highlighted in the respective colour. The left hand pane shows the resources loaded into GATE: LRs consisting of a corpus and the document it contains; PRs consisting of the different components such as the tokeniser, gazetteer, etc., being used; and an

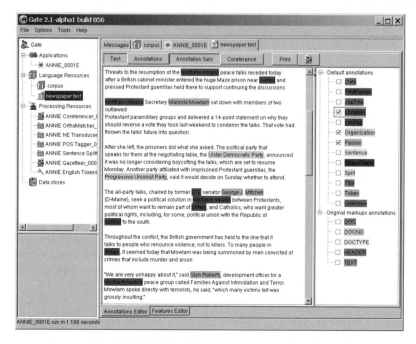

Figure 4.1   GATE GUI with document annotated by ANNIE.

application which combines the necessary PRs for the task and runs them on the desired corpus. The main window shows the text highlighted with the annotations, and can also show further details of each annotation, allowing the user to edit these annotations manually. The right hand pane shows the annotation types, arranged into sets for convenience, and can also show other information such as co-reference, if such a component has been run.

4.4   Text analysis components

In this section we describe a number of PRs which are general enough to start development of text-based BI applications.

4.4.1   Document structure identification

The first stage in any application in GATE is the structural analysis of the document to be processed. This involves identifying the format of the document (word, PDF, plain text, HTML, etc.), stripping off any formatting which is not part of the text, such as HTML tags or JavaScript, and making the document readable to the human eye. Any tags stripped off at this stage are kept and converted into annotations on the document so that they can be used for later processing if applicable. GATE currently recognises the following document formats: plain text, email, HTML, XML, SGML, Microsoft Word, RTF and PDF (though the last three may be slightly error prone depending on the complexity of the document). Document structure identification is performed automatically when a text is loaded into GATE.

4.4.2   Tokenisation

The tokeniser splits the text into very simple tokens such as numbers, punctuation and words of different types. For example, we distinguish between words in upper case and lower case, and between certain types of punctuation. The aim is to limit the work of the tokeniser to enable greater flexibility by placing the burden on the grammar rules, which are more adaptable. The default tokeniser is both domain and language independent, though minor modifications may be useful for specific languages.

The English Tokeniser is a PR that comprises a default Unicode tokeniser and a JAPE transducer. The transducer has the role of adapting the generic output of the tokeniser to the requirements of the English part-of-speech tagger (see below). One such adaptation is to join together into one token constructs like '30s', 'Cause', 'em', 'N', 'S', 's', 'T', 'd', 'll', 'm', 're', 'til', 've', etc. Another task of the JAPE transducer is to convert negative constructs like 'don't' from three tokens ('don', ''' and 't') into two tokens ('do' and 'n't').

4.4.3   Sentence identification

The sentence splitter is a cascade of finite state transducers which segments the text into sentences. This module is required for the tagger. Each sentence is annotated with the type Sentence. Each sentence break (such as a full stop) is also given a 'Split' annotation. This has several possible types: '.', 'punctuation', 'CR' (a line break) or 'multi' (a series of punctuation marks such as '?!?!'). The sentence splitter is domain and application independent, and to a certain extent language independent, though it relies (for English) on a small lexicon of common abbreviations to distinguish between full stops marking these from full stops marking ends of sentences.

4.4.4   Part of speech tagging

The Part of Speech (POS) tagger is a modified version of the Brill tagger (Brill, 1994), which produces a part-of-speech tag as an annotation on each word or symbol. The tagger uses a default lexicon and ruleset which is the result of training on a large corpus taken from the *Wall Street Journal*. Both of these can be modified manually if necessary. While the POS tagger is clearly language dependent, experiments with other languages (see e.g. Maynard *et al.*, 2003b) have shown that by simply replacing the English lexicon with an appropriate lexicon for the language in question, reasonable results can be obtained, not only for Western languages with similar word order and case marking to that of English, but even for languages such as Hindi, with no further adaptation.

4.4.5   Morphological analysis

Morphological analysis is the process of transforming a word into a string of its morphological components. For example, a verb such as *gives* could be transformed by morphological analysis into the root *give* and the suffix *s* which indicates the third-person singular (he, she, it). In GATE, this process is carried out by a PR which enriches the token annotation with two features, *root* (for the word root) and *affix* (for the word ending). No analysis of prefixes or infixes is carried out. The analysis is carried out on nouns and verbs; other word forms are normalised by transforming the word into all lower case (e.g. *THE* into *the*).

4.4.6   Stemming

Stemming is the process of removing suffixes from words by automatic means. It is a useful process in information retrieval because it helps reduce the number of terms used to represent documents. It can also help to increase the performance of the information retrieval system. Sometimes words with the same stem would have similar meanings – for example, *connect, connected, connecting, connection, connections* could be reduced to the same stem *connect*. In an information retrieval system using stemming, a search for a document containing the word *connected* would therefore yield documents containing words such as *connection*, which can be of some advantage. In GATE, stemming of English texts is carried out with the well-known Porter algorithm. GATE also provides stemmers in other languages.

4.4.7   Gazetteer lookup

Gazetteer lookup consists of a set of gazetteer lists which are run over the text as finite state machines, and which create annotations on the text. The gazetteer lists are plain text files, with one entry per line. Each list represents a set of names, such as names of cities, organisations, days of the week, etc. Gazetteer lists can be set at runtime to be either case sensitive or case insensitive and can also be set either to match whole words only or to match parts of words also. An index file is used to access these lists: for each list, a major type is specified and, optionally, a minor type. It is also possible to include a language in the same way, where lists for different languages are used. These lists are compiled into finite state machines. Any text matched by these machines will be annotated as type 'Lookup' with features specifying the values of the major and minor types defined for that file in the index file.

4.4.8   Name recognition

Name recognition is performed by the semantic tagger component, which consists of a set of grammars based on the JAPE language. The grammars contain rules which act on annotations assigned to the text previously, in order to produce outputs of annotated entities. Each grammar set contains a series of JAPE grammars run sequentially, such that annotations created by one grammar may be used by a following grammar. This is very important for ambiguity resolution between entity types.

In the simple JAPE rule example below, the pattern described will be awarded an annotation of type 'Location'. This annotation will have the attribute 'kind', with value 'unknown', and the attribute 'rule', with value 'GazLocation'. (The purpose of the 'rule' attribute is simply for debugging.)

```
Rule: GazLocation
({Lookup.majorType == location})
:loc -->
:loc.Location = {kind="unknown", rule=GazLocation}
```

Most grammar rules use a combination of gazetteer lookup and POS tags, though they may include any kind of annotation such as token type, orthographic information, token length or previous entity annotations found in an earlier phase. Feature information can also be passed from a matched annotation using a more complex kind of rule involving Java code on the right hand side (RHS) of the rule. For example, the minor type from a gazetteer lookup can be percolated into the new annotation, such that we can retain information in the final entity annotation about whether a person is male or female, or classification information about locations.

JAPE is also aware of any ontology being used (see below), making the mapping of entities into ontological classes possible during entity recognition.

4.4.9   Orthographic co-reference

The orthomatcher module detects orthographic co-reference between named entities in the text (e.g. James Smith and Mr Smith). It has a set of hand-crafted rules, some of which apply for all types of entities, while others apply only for specific types, such as persons or organisations. A similar module exists for identifying pronominal co-reference (e.g. between 'he' and 'Mr Smith'). The orthomatcher is described more fully in Dimitrov et al. (2002).

The orthomatcher is also used to classify unknown proper names and thereby improve the name recognition process. During the named entity recognition phase, some proper nouns are identified but are simply annotated as Unknown, because it is not clear from the information available whether they should be classified as an entity and, if so, what type of entity they represent. A good example of this is a surname appearing on its own without a title or first name, or any other kind of indicator such as a conjunction with another name, or context such as a job title. The orthomatcher tries to match Unknown annotations with existing annotations, according to the same rules as before. If a match is found, the annotation type is changed from Unknown to the type of the matching annotation, and any relevant features such as gender of a Person are also added to match. Two Unknown annotations cannot be matched with each other. Also, no annotation apart from an Unknown one can be matched with an existing annotation of a different type, for example a Person can never be matched with an Organisation, even if the two strings are identical, and its annotation type cannot be changed by the orthomatcher. So, for example, 'Smith' occurring on its own in the text might be annotated by the JAPE transducer as Unknown, but if 'Mr Smith' is also found in the text (and annotated as a Person), the orthomatcher will find a match between these two strings, and will change the Unknown annotation into a Person one.

4.4.10   Parsing

Parsing is the process of assigning to a sentence a syntactic analysis based on a grammar of the language. Several parsing algorithms – or parsers – are provided in GATE as PRs. One such parser is SUPPLE (Gaizauskas et al., 2005). SUPPLE is a bottom-up parser that constructs syntax trees and logical forms for English sentences. The parser is complete in the sense that every analysis licensed by the grammar is produced. In the current version only the 'best' parse is selected at the end of the parsing process. The English grammar is implemented as an attribute–value context-free grammar which consists of sub-grammars for Noun Phrases (NPs), Verb Phrases (VPs), Prepositional Phrases (PPs), Relative phrases (Rs) and Sentences (Ss). The semantics associated with each grammar rule allow the parser to produce logical forms composed of unary predicates to denote entities and events (e.g. chase(e1), run(e2)) and binary predicates for properties (e.g. lsubj(e1,e2)). Constants (e.g. e1, e2) are used to represent entity and event identifiers. The GATE SUPPLE Wrapper stores syntactic information produced by the parser in the GATE document in the form of 'parse' annotations containing a bracketed representation of the parse, and 'semantics' annotations that contain the logical forms produced by the parser. Other parsers such as Minipar and RASP are also available in GATE.

4.5   Ontology support

As discussed in Chapter 3, there are different notions in the literature and in the different research communities on what ontologies are – or should be. The most widely used one is: 'An ontology is an explicit specification of a shared conceptualisation' (Gruber, 1995). The Semantic Web (Fensel et al., 2002) and related developments have ontologies as their foundation and have consequently increased the need for text analysis applications that use ontologies (Fensel, 2001). Text analysis algorithms use ontologies in order to obtain a formal representation of and reason about their domain (Saggion et al., 2007; Handschuh et al., 2002). In addition, text analysis methods can be used to populate an ontology with new instances discovered in texts or even to construct an ontology automatically. In other words, text analysis can have ontologies as target output and also use them as an input knowledge source. GATE provides support for importing, accessing and visualising ontologies as a new type of linguistic resource.

The proliferation of ontology formats for the Semantic Web, for example OWL (Dean *et al.*, 2004) and RDF Schema or RDF(S) (Brickley and Guha, 2004), means that text analysis applications need to deal with these different formats. In order to avoid the cost of having to parse and represent ontologies in each of these formats in each text analysis system, GATE offers a common object-oriented model of ontologies with a unified API. The core ontology functionality is provided through the integration of OWLIM, a high-performance semantic repository developed by Ontotext (Kiryakov, 2006). OWLIM has an inference engine (TREE) to perform RDF(S) and OWL reasoning. The most expressive language supported is a combination of limited OWL Lite and unconstrained RDF(S). OWLIM offers configurable reasoning support and performance. In the 'standard' version of OWLIM (referred to as SwiftOWLIM) reasoning and query evaluation are performed in memory, while a reliable persistence strategy assures data preservation, consistency and integrity.

This approach has well-proven benefits, because it enables each analysis component to use this format-independent ontology model, thus making it immune to changes in the underlining formats. If a new ontology language/format needs to be supported, the text analysis modules will automatically start using it, due to the seamless format conversion provided by GATE. From a developer's perspective the advantage is that they only need to learn one API and model, rather than having to learn many different and rather idiosyncratic ontology formats.Since OWL and RDF(S) have different expressive powers, the GATE ontology model consists of a class hierarchy with a growing level of expressivity. At the top is a taxonomy class which is capable of representing taxonomies of concepts, instances and inheritance between them.

At the next level is an ontology class which can represent also properties: that is, relate concepts to other concepts or instances. Properties can have cardinality restrictions and be symmetric, transitive, functional, etc. There are also methods providing access to their sub- and super-properties and inverse properties. The property model distinguishes between object (relating two concepts) and datatype properties (relating a concept and a datatype such as a string or number).

The expressivity of this ontology model is aimed at being broadly equivalent to OWL Lite. Any information outside the GATE model is ignored. When reading RDF(S), which is less expressive than OWL Lite, GATE only instantiates the information provided by that model. If the API is used to access one of these unsupported features, it returns empty values.GATE also offers basic ontology visualisation and editing support (see Figure 4.2), because it is needed for carrying out manual semantic annotation (i.e. annotating texts with respect to an ontology); it also facilitates the development and testing of analysis modules that use ontologies (e.g. gazetteers and JAPE-based modules).

A very important step in the development of semantic annotation systems is the creation of a corpus for performance evaluation. For this purpose, GATE has an Ontology Annotation Tool (OAT) (see Figure 4.3) which allows users to

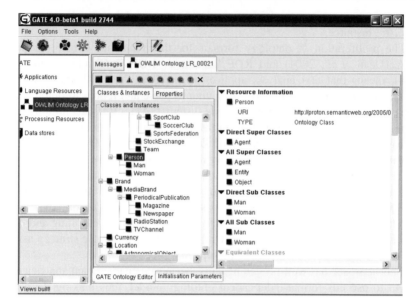

Figure 4.2   GATE ontology visualisation and editing support.

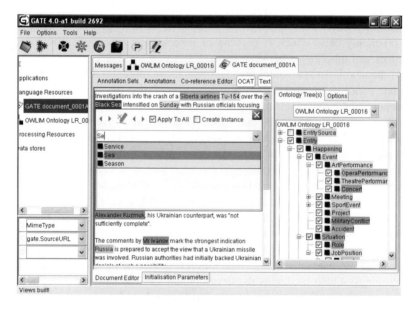

Figure 4.3   GATE ontology annotation tool.

annotate texts with references to ontology classes, instances and properties. If the required information is not already present in the ontology, OAT allows the user to extend the ontology with a new instance at the same time as annotating the text.

The ontology support is currently being used in many projects that develop Ontology-Based Information Extraction (OBIE) to populate ontologies with instances derived from texts and also to annotate texts with mentions of ontology classes, instances and properties (Saggion *et al.*, 2007). The OBIE systems produce annotations referring to the concepts and instances in the ontology, based on their URIs (Uniform Resource Identifiers).The JAPE pattern matching engine was also extended to provide access to ontologies on the right hand side of JAPE rules. This allows JAPE rules to add new information to ontologies (e.g. instances) or to consult them (e.g. to obtain semantic distance between concepts). The JAPE finite state machine also takes into account subsumption relations from the ontology when matching on the left hand side. So, for example, a rule might look for an organisation followed by a location, in order to create the `locatedAt` relationship between them. If JAPE takes into account subsumption, then the rule will automatically match all subclasses of **Organisation** in the ontology (e.g. **Company**, **GovernmentOrg**).

4.6   Ontology-based information extraction

OBIE has two main purposes: automatic document annotation and automatic ontology population.

Automatic annotation consists of annotating mentions of instances in a text with their corresponding classes (concepts) from an ontology (e.g. to annotate 'John Smith' with the concept 'Person'). For the purposes of automatic annotation, the OBIE process needs to perform the following two tasks:

1. Identify mentions in the text, using the classes from the ontology instead of the flat list of types in 'traditional' information extraction systems.

2. Perform disambiguation (e.g. if we find two occurrences of 'John Smith' in the text, where one refers to a person and the other to the name of the beer, the first might be annotated with the concept 'Person' and the second with the concept 'Beer').

It may also perform a third task: identifying relations between instances.

Ontology population involves automatically generating new instances in a given ontology from a data source. It links unique mentions of instances in the text to instances of concepts in the ontology. This component is similar in function to the automatic annotation component. However, it requires not only that instances are disambiguated, but also that co-referring instances are identified. For example, if we find two occurrences of 'John Smith' in the text, where one refers to a person and the other to the name of the beer, then our system should add the first as an instance of the concept 'Person' and the second as an instance of the concept 'Beer'. On the other hand, if we find an occurrence of 'John Smith' in the text and an occurrence of 'Mr Smith', the system must identify whether they are referring to the same person or to two different people (or even that one if referring to the beer and the other to a person), and if they are co-referring, then only one instance should be added to the ontology.

For the development of ontology-based IE systems GATE also provides a tool called OntoRootGazetteer, which creates a gazetteer lookup process from the ontology of the domain, making it possible to match instances, class names and labels found in a document. This component is particularly useful for the development of OBIE systems.

4.6.1   An example application: market scan

The tools presented above have been used in the MUSING project to implement an OBIE application for a typical internationalisation problem: that is, to extract relevant geographical, social, institutional, political and economic information from a number of documents about various geographical regions. In the real-life case studied, the tests were focusing on the Indian subcontinent (MUSING, 2006). The application consists of an automatic tool for identifying, with priorities, regions in India most suited for various forms of collaboration with European SMEs. Considerations are based on matching areas of interest and compatibility in business and technological capabilities. The service offered consists of filling in a questionnaire with details on the SME and the collaboration sought. Based on this input, the application performs OBIE based on two ontologies: an ontology indicator (containing target concepts) and an ontology of Indian regions (containing the target regions for the target concepts). The application finds specific 'mentions' of indicators and data associated with them, for example the average population of an area, but also finds sentences containing useful information related to one of the indicators. For example, the following sentence is an example of political instability:

> In the 1980s the Brahmaputra valley saw a six-year Assam Agitation
> that began non-violently but became increasingly violent.

This is annotated simply as an instance of the PoliticalInstability concept. Similarly, the following sentence is annotated as an example of the MarketSize concept:

> The per capita income of Assam was higher than the national average
> soon after Indian Independence.

For each region, the application finds values for the following indicators about regions: population, density of population, surface, number of districts, forestation, literacy rates, educational institutions, political stability indicators, etc. The target Indian region ontology simply classifies all the Indian provinces, such as Assam, Bihar, etc., as subRegionOf India. This enables us to recognise each mention of an Indian province when it occurs in the text. The Indicator ontology specifies a number of indicator keywords which can be used for classification purposes. For example, it has several keywords representing economic stability issues. This is represented by the concept EconomicStabilityIndicator, which contains a number of instances (indicators) such as ExchangeRateVolatility, ImportValue and ExportValue. Similarly, the concept MarketSizeIndicator has instances such as PopulationDensity, NumberOfDistricts, SurfaceArea, and so on.

The extraction application consists of a number of processing resources: general linguistic pre-processing (tokenisation, POS tagging, morphological analysis, named entity recognition), and document-specific pre-processing to extract some relevant information based on the document structure. The main OBIE processing involves the actual OBIE part. It performs ontology lookup and then uses grammar rules to find the relevant concepts, instances and relations in the text. The main module is the OntoRootGazetteer module, which looks for keywords from the ontology in the text. These could be represented in the ontology and/or text as any morphological variant (the morphological analyser used in the linguistic pre-processing phase finds the root of the word in each case, and the results of this are used in order to perform the matching between ontology and text at the root level of the token) after a gazetteer lookup which looks for certain other indicator words not found in the ontology. Finally, JAPE grammars are used to perform recognition of the relevant mentions. This includes both looking for specific items like Population sizes and names of universities, and also recognition of sentences containing indicator keywords (e.g. sentences mentioning 'forests' and so on). These grammars make use both of the regular gazetteers and the lookups from the ontology (via the OntoRootGazetteer lookup). They produce Mention annotations with a set of features, which are then used to create the RDF in the mapping phase.

Figure 4.4 shows the results of annotation of some of the Indicators in the text. Each Indicator is annotated with the type 'Mention' which contains various features and values describing the kind of Indicator and other information which will be used to create the final RDF output. In this screenshot, we see some highlighted sentences about forestation and rainfall. These are related to the Resource Indicator – more specifically, to the Forest Indicator.

4.7   Evaluation

GATE provides a variety of tools for automatic evaluation. A tool called Annotation Diff compares two sets of annotations within a document using precision and recall measures. A corpus Quality Assurance tool extends Annotation Diff to an entire corpus or collection of documents. Other tools provide support for computation of other measures (e.g. inter-annotator agreement, kappa). These tools are particularly useful not just as a final measure of performance, but as a tool to aid

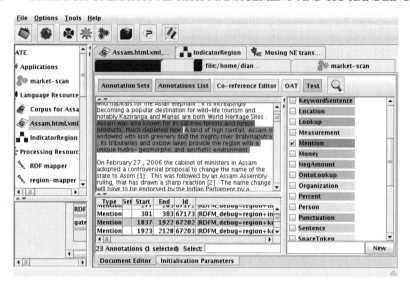

Figure 4.4   Annotation of Indicators in free text.

system development by tracking progress and evaluating the impact of changes as they are made. Applications include evaluating the success of a machine learning or language engineering application by comparing its results to a gold standard and also comparing annotations prepared by two human annotators to each other to ensure that the annotations are reliable.

We make use of traditional metrics used in IE (Chinchor *et al.*, 1993): precision, recall and *F*-measure. Precision measures the number of correctly identified items as a percentage of the number of items identified. It measures how many of the items that the system identified were actually correct, regardless of whether it also failed to retrieve correct items. The higher the precision, the better the system is at ensuring that what is identified is correct. Recall measures the number of correctly identified items as a percentage of the total number of correct items measuring how many of the items that should have been identified actually were identified. The higher the recall rate, the better the system is at not missing correct items. The *F*-measure is often used in conjunction with Precision and Recall, as a weighted average of the two – usually an application requires a balance between Precision and Recall.

4.8   Summary

The ubiquitous presence of online textual information on the Web calls for appropriate resources and tools for content analysis of unstructured sources. Without resources that extract and transform textual information into a standard format, the work of business intelligence analysts would be impossible. This chapter has briefly described a number of natural language processing resources for the development of text analysis and information extraction applications in business intelligence. The chapter has described a number of processes usually applied in the field of information extraction, illustrating them with the GATE system, an open platform implemented in Java used worldwide in various projects, including the business intelligence MUSING project (MUSING, 2006).

References

Appelt, D.E., Hobbs, J.R., Bear, J., Israel, D., Kameyama, M., Martin, D. Myers, K. and Tyson, M. (1993) SRI International FASTUS System: MUC-6 Test Results and Analysis. *Proceedings of the 6th Conference on Message Understanding*, Columbia, Maryland.

Bontcheva, K., Tablan, V., Maynard, D. and Cunningham, H. (2004) Evolving GATE to Meet New Challenges in Language Engineering. *Natural Language Engineering*, 10, 3/4, pp. 349–373.

Brickley, D. and Guha, R.V. (2004) RDF Vocabulary Description Language 1.0: RDF Schema. W3C Recommendation. http://www.w3.org/TR/rdf-schema (6 March 2010).

Brill, E. (1994) Some Advances in Transformation-Based Part of Speech Tagging. *Proceedings of the 1994 AAAI Conference*, Seattle, pp. 722–727.

Chinchor, N., Hirschman, L. and Lewis, D.D. (1993) Evaluating Message Understanding Systems: An Analysis of the Third Message Understanding Conference (MUC-3). *Computational Linguistics*, 19, 3, pp. 409–449.

Chung, W., Chen, H. and Nunamaker, J.F. (2003) Business Intelligence Explorer: A Knowledge Map Framework for Discovering Business Intelligence on the Web. *Hawaii 2003 International Conference on System Sciences*, Big Island, Hawaii.

Ciravegna, F. (2001) LP2: An Adaptive Algorithm for Information Extraction from Web-related Texts. *Proceedings of the IJCAI-2001 Workshop on Adaptive Text Extraction and Mining*.

Cunningham, H., Maynard, D. and Tablan, V. (2000) JAPE: A Java Annotation Patterns Engine (Second Edition). Research Memorandum CS–00–10, Department of Computer Science, University of Sheffield.

Cunningham, H., Maynard, D., Bontcheva, K. and Tablan, V. (2002) GATE: A Framework and Graphical Development Environment for Robust NLP Tools and Applications. *Proceedings of the 40th Anniversary Meeting of the Association for Computational Linguistics (ACL'02)*, Philadelphia.

Dean, M., Schreiber, G., Bechhofer, G., van Harmelen, F., Hendler, J., Horrocks, I., McGuinness, D.L., Patel-Schneider, P. and Stein, L.A. (2004) OWL Web Ontology Language Reference. W3C Recommendation. http://www.w3.org/TR/owl-ref/ (accessed 21 May 2010).

Dimitrov, M., Bontcheva, K., Cunningham, H. and Maynard, D. (2002) A Light-weight Approach to Conference Resolution for Named Entities in Text. *Proceedings of the Fourth Discourse Anaphora and Anaphor Resolution Colloquium (DAARC)*, Lisbon.

Fensel, D. (2001) *Ontologies: Silver Bullet for Knowledge Management and Electronic Commerce*. Springer-Verlag, Berlin.

Fensel, D., Wahlster, W. and Lieberman, H. (Eds) (2002) *Spinning the Semantic Web: Bringing the World Wide Web to Its Full Potential*. MIT Press, Cambridge, MA.

Gaizauskas, R., Hepple, M., Saggion, H., Greenwood, M.A. and Humphreys, K. (2005) SUPPLE: A Practical Parser for Natural Language Engineering Applications. *Proceedings of the International Workshop on Parsing Technologies*.

Gruber, T. (1995) Toward Principles for the Design of Ontologies Used for Knowledge Sharing. *International Journal of Human and Computer Studies*, 43, 5/6, pp. 907–928.

Handschuh, S., Staab, S. and Ciravegna, F. (2002) S-CREAM Semi-automatic CREAtion of Metadata. *13th International Conference on Knowledge Engineering and Knowledge Management (EKAW02)*, Siguenza, Spain, pp. 358–372.

Isozaki, H. and Kazawa, H. (2002) Efficient Support Vector Classifiers for Named Entity Recognition. *COLING 2002: Proceedings of the 19th International Conference on Computational Linguistics*, Taipei, Taiwan.

Kiryakov, A. (2006) OWLIM: Balancing Between Scalable Repository and Lightweight Reasoner Developer's Track of WWW2006, *15th International World Wide Web Conference*, Edinburgh.

Lafferty, J., McCallum, A. and Pereira, F. (2003) Conditional Random Fields: Probabilistic Models for Segmenting and Labelling Sequence Data. *Proceedings of the 18th International Conference on Machine Learning*, Williamstown, Massachusetts.

Leek, T.R. (1997) Information Extraction using Hidden Markov Models. Master's Thesis, UC San Diego.

Maynard, D., Bontcheva, K. and Cunningham, H. (2003a) Towards a Semantic Extraction of Named Entities. *Recent Advances in Natural Language Processing*, Bulgaria.

Maynard, D., Tablan, V., Bontcheva, K., Cunningham, H. and Wilks, Y. (2003b) Rapid Customisation of an Information Extraction System for Surprise Languages. Special issue of *ACM Transactions on Asian Language Information Processing: Rapid Development of Language Capabilities: The Surprise Languages*.

MUSING (2006) IST-FP6 27097. http://www.musing.eu (accessed 21 May 2010).

Riloff, E. (1993) Automatically Constructing a Dictionary for Information Extraction Tasks. *Proceedings of the Eleventh Annual Conference on Artificial Intelligence*, Detroit, pp. 811–816.

Saggion, H., Funk, A., Maynard, D. and Bontcheva, K. (2007) Ontology-based Information Extraction for Business Applications. *Proceedings of the 6th International Semantic Web Conference* (ISWC 2007), Busan, Korea.

Yangarber, R., Grishman, R., Tapanainen, P. and Huttunen, S. (2000) Unsupervised Discovery of Scenario-level Patterns for Information Extraction. *Proceedings of ANLP-NAACL'00*, Seattle.

5

A case study of ETL for operational risks

Valerio Grossi and Andrea Romei

5.1 Introduction

In accordance with the International Convergence of Capital Measurement and Capital Standards, known as Basel II (Basel Committee on Banking Supervision, 2006), *operational risk* is defined as 'the risk of loss resulting from inadequate or failed internal processes, people and systems, or from external events'. In particular, in our specific context, the term ICT operational risk is adopted to consider events of business disruption related to system failures (e.g. hardware or software failures).

As stated in Azvine *et al.* (2007), operational risk management (OpR) can be considered 'as a number of possibly overlapping components such as strategic, operational, financial and technology-oriented'. In other words, operational risk is a function of the complexity of the business and the environment that the business operates in. The higher the business, the more such complexities increase with the aim of producing operational risk indicators. A critical measure of such a complexity is given by answering questions, such as 'how to pull in the right operational data?' or 'how to automate the collection, cleansing, aggregation, correlation and analysis processes in order to perform the operational risk management both more effectively and more efficiently?'

Operational Risk Management: A Practical Approach to Intelligent Data Analysis   Edited by Ron S. Kenett and Yossi Raanan   © 2011 John Wiley & Sons, Ltd

As a consequence, the more complexity increases, the higher the need for integrating both internal and external disparate data sources, and filtering external data according to internal rules and definitions to eliminate irrelevant data. This statement is especially valid when internal loss data is insufficient for effective risk indicator calculations. In these cases, techniques for merging and integrating data become a critical aspect.

The class of tools responsible for these tasks is known as *extract, transform and load (ETL)*. They are an important component of business analysis, since they represent the first step of a business process in which the data is actually gathered and loaded into a data warehouse. As far as OpR is concerned, the generic functionalities of an ETL process may include: (1) the identification of relevant information for risk assessment at the source side; (2) the extraction of this information; (3) the customization and integration of the information from multiple sources into a common format; (4) the cleaning of the resulting data set, on the basis of both database and business OpR rules; (5) the propagation of the data to the data warehouse for operational risk purposes and for the (automatic or semi-automatic) generation of risk indicators.

This chapter presents an application of typical ETL processes used to carry out the analysis of causes of failure by merging the data available from different and heterogeneous sources. The application was developed in the context of the MUSING project (MUSING, 2006). The specific aims of the study were to enable risk identification and assessment of operational risks, and then to provide some guidelines for risk monitoring and mitigation. More specifically, we present a case study in OpR in the context of a network of private branch exchanges (PBXs), managed by a virtual network operator (VNO). Given a set of technical interventions and PBX monitoring status, our aim is to merge these two kinds of data to enable an analysis of correlations between some specific problems and well-defined sequences of alarms. We highlight the data available, the problems encountered during the data merging, and finally the solution proposed and implemented by means of an ETL tool. Finally, information related to the business activities of a specific customer is integrated in the merged data, in order to enable the analyst to consider even business indicators provided by the MUSING platform. This information enables the analyst to calibrate different policies for managing operational risks by also considering business factors related to the customer under analysis.

The chapter is organized as follows. Section 5.2 provides an overview of the current state of the art in ETL techniques. This section includes a brief description of Pentaho, an open source ETL suite adopted in our case study (Pentaho, 2010). Section 5.3 describes the scenario of our application, its integration in the MUS-ING architecture, as well as a detailed description of the data sources available and the data merging flow. The section introduces several relevant issues of data merging. Section 5.4 presents the main ETL tasks related to the data merging case study from a physical point of view. The focus in mainly on the solution of the problem and on the results obtained. Finally, Section 5.5 summarizes the chapter.

5.2   ETL (Extract, Transform and Load)

ETL is a process for extracting and transforming data from several sources. It enforces quality and data consistency, and guarantees a common structure in the presence of different and heterogeneous inputs. An ETL system removes mistakes and corrects missing data, converts datatypes, unifies the data structures from multiple sources to be used together and filters irrelevant data.

However, building an ETL system is a rather complex task. As stated in Kimball and Caserta (2004), it typically covers more than 70% of the resources needed for the implementation and maintenance of a typical data warehouse. On the other hand, if the application of ETL techniques is correct and precise, the integrated data source can be easily used: a programmer can quickly build new applications over it, an analyst can take more accurate decisions, a statistician can directly visualize reports and, finally, a data mining expert can use the data to extract mining models.

Figure 5.1 depicts a typical application of an ETL system. As shown, the *data warehouse* is a central integrated database, containing data from operational sources in an organization, such as relational and object-oriented DBMSs, flat files, XML files, spreadsheets, XML documents, and so on. It may gather manual inputs from users, determining criteria and parameters for grouping or classifying records. The database contains structured data for query analysis and can be accessed by users.

More importantly, the data warehouse can be created or updated at any time, with minimum disruption of operational systems. This task is ensured by an

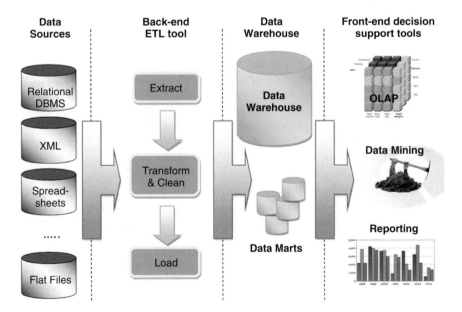

Figure 5.1   A typical data-warehouse-based business intelligence architecture.

ETL process. Preparing the data involves acquiring data from heterogeneous sources (Extract). It is then cleaned, validated and transformed into 'information' (Transform & Clean). Stages at this step range from simple data mappings from one table to another (with renaming of attributes and type conversions), to joining data from two or more data paths or to complex splitting of data into multiple output paths that depend on input conditions. Finally, data is aggregated and prepared to be loaded into the data warehouse for business analysis (Load). Moreover, when only a portion of detailed data is required, it is convenient to use a *data mart*, which contains data focused on a given subject and data frequently accessed or summarized.

From Figure 5.1 it is clear that ETL is a central task in data warehouse system architectures with business intelligence purposes, since the output of the data merging becomes the input of important business operations, such as online analytical processing (OLAP), data mining and data visualization.

5.2.1   Related work

To the best of our knowledge, the MUSING case study is the first approach to solve problems related to OpR by means of ETL techniques. In this section, the work related to ETL is considered. We distinguish between work covering the ETL conceptual model and work focusing on the optimization of ETL workflow execution.

5.2.2   Modeling the conceptual ETL work

To handle the conceptual level of an ETL application, a simple, but sufficiently expressive, model is specifically required for the early stages of data warehouse design.

In the last few years, several approaches have focused on the conceptual part of the design and development of ETL scenarios (Kimball and Caserta, 2004; Kof, 2005; Luján-Mora, Vassiliadis and Trujillo, 2004; Vassiliadis, Simitsis and Skiadopoulos, 2002). Moreover, several commercial tools already exist and all major DBMS vendors provide powerful graphical design and representation tools for ETL processes, namely IBM WebSphere DataStage, Informatica Power-Center, Microsoft Data Transformation Services, Oracle Warehouse Builder and Wisnesky *et al.* (2008), to cite but a few. In many cases, commercial and open source ETL tools provide specifications in some proprietary formats (e.g. based on XML) and decreasing interoperability features, which are typically required in complex scenarios where several software packages are involved in monitoring and managing data. Generally, as stated in Simitsis, Skoutas and Castellanos (2008), the proposed solutions require 'extra knowledge' from the user, in order to become familiar with the symbolism used. The lack of a uniform modeling technique is a crucial issue in data warehousing. The work proposed by Simitsis *et al.* tries to integrate the emerging usage of ontology with natural language techniques, to facilitate and clarify the conceptual ETL design by the production of requirement reports for ETL processes (Simitsis *et al.*, 2006).

Although ETL processes are quite important in data integration projects, little work on ETL as a meta-level has been conducted in the database research community. In particular, Albrecht and Naumann (2008), Bernstein (2001) and Melnik (2004) propose a generic approach to enable a flexible reuse, optimization and rapid development of ETL processes. As for SQL, the aim of these papers is to produce a standard language and representation for treating ETL processes in a tool-independent representation.

5.2.3   Modeling the execution of ETL

An ETL-based workflow usually follows a quite complex design, which applies complex operations over (typically) large volumes of data. Since an ETL job must be completed in a specific time window, it is necessary to optimize its execution time as much as possible.

The most extensive study in this direction is that by Simitsis, Vassiliadis and Sellis (2005), who propose a multi-level workflow model that can be used to express ETL jobs. Such jobs can be subsequently analyzed and optimized via logical inference rules. A system called Orchid uses a simplified version of this common model tailored to deal with mappings, as stated in Dessloch *et al.* (2008). In detail, it compiles real ETL jobs into a common model and deploys the resulting abstract model instance into a valid job in an ETL system or other target platform.

Another appreciable contribution that uses a logical model for workflow optimization is reported in Sellis and Simitsis (2007). As stated by the authors, their main results include: (1) a novel conceptual model for tracing inter-attribute relationships and the respective ETL transformations, along with an attempt to use ontology-based mechanisms to capture semi-automatically the semantics and relationships among the various sources; (2) a novel logical model for the representation of ETL workflows with the characteristics of genericity and customization; (3) the semi-automatic transition from the conceptual to the logical model for ETL workflows; and finally (4) the tuning of an ETL workflow for optimization of the execution order of its operations.

5.2.4   Pentaho data integration

As mentioned above, several commercial tools support the ETL process. These systems adopt a visual metaphor; the process is modeled as a workflow of transformation tasks with data flowing from data sources to data destinations. Most vendors implement their own core set of operators for ETL, and provide a proprietary GUI to create a data flow. From the database perspective, the absence of a standard implies that there is no common model for ETL. Therefore, in the plethora of open source solutions, only those that offer the largest operator repository, a user-friendly GUI to compose flows and versatile mechanisms to adapt to the various use cases are bound to emerge. An established leader in ETL solutions is the open source Pentaho Data Integration (Pentaho, 2010).

As a typical ETL suite, Pentaho is able to extract data from various sources, such as text files, relational and XML data, transforming and loading them into a data warehouse. Using the graphical design tool of the Pentaho Data Integration, we can graphically create the ETL processes. Therefore, as a visual product, we have user-friendly access operators for data source access, transformation and loading, as well as debug support and similar functionalities. As an ETL product, we have generality, due to a wide range of data source/destination types and transformation tasks.

More specifically, a transformation is represented by a graph, in which we can have one or more data inputs, several steps for handling data and finally nodes to store the data. Every operator, represented by a node in the graph, is connected by a hop indicating the stream of data from one node to another. From a logical point of view, every operator performs an operation on the data, computing new fields, aggregations and joins. In addition, user-defined data modification can also be introduced by means of JavaScript programs. Pentaho Data Integration can also interact with web services by using standard WSDL (Web Service Description Language) descriptions to define the number and types of inputs (and outputs) required by the service.

Pentaho transformations can be used to define a job represented with the same transformation formalism. Clearly, in a job definition every graph node is a transformation, while a hop represents the link between two job entries. In particular, a job enables control of the transformation execution flow, for example executing the next job entry unconditionally, or selecting a particular job branch after a successful (or failed) node execution. In addition, with the Pentaho Data Integration suite we can execute and debug the defined transformation, as well as execute it on only a small portion of data. Transformations and jobs are stored using a proprietary-defined XML file.

5.3   Case study specification

The case study aims at developing a tool capable of merging data from text logs, technician interventions as well as financial information on business enterprises. It concerns the management of a network of PBXs by a VNO telecommunication service provider. This work needs to be automated, since the application developed should be able to access, pre-process and merge the available data without manual intervention.

The merged data obtained can be subsequently used to study and design (customer-)specific policies for OpR.

5.3.1   Application scenario

The customers of the VNO are SMEs and large enterprises requiring both voice and data lines at their premises, at different contractually agreed qualities of service. The customers outsource the maintenance of the PBXs and the actual management of the communication services to the VNO.

When a malfunction occurs, customers refer to the VNO's call center, which can act remotely on the PBX, for example rebooting the system. Unfortunately, if the problem cannot be solved remotely, as in the case of a hardware failure, a technician is sent to the site. It is clear that the latter situation is more expensive than remote intervention by the call-center operator.

Both call-center contacts and technician reports are logged in the VNO customer relationship management database. In particular, a PBX is doubly redundant: that is, it actually consists of two independent communication apparatuses and self-monitoring software. Automatic alarms produced by the equipment are recorded in the PBX system log. Call-center operators can access the system log to control the status of the PBX. In addition, centralized monitoring software collects system logs from all the installed PBXs on a regular basis in a round-robin fashion.

Among the operational risk events, PBX malfunctioning may have a different impact on the VNO. At one extreme, the call-center operator can immediately solve the problem. At the other, a technician intervention may be required, with the customer's offices inoperative in the meantime, and the risk that the contractual SLA or general quality of service might not be guaranteed. To record the impact of a malfunction, the severity level of the problem occurring is evaluated and documented by the technician.

The customers of the VNO are also business enterprises, such as banks, industries, governments, insurance agencies, and so on. The aim of the merging execution is to produce data integration to provide a single data source comprising specific financial information, such as details of fixed assets, profits, goodwill (reflecting the good relationship of a business enterprise with its customers). This information is gathered from company balance sheets.

In this context, our case study uses ETL to achieve a merged database of the available sources of data, including customer type, financial information, call-center logs, technicians' reports and PBX system logs.

5.3.2   Data sources

In this section, we provide a description of the data sources available in our case study. All the data is described at the logical level concentrating on only the most important features influencing the data merging.

5.3.2.1   TECH

Data has been provided by a leading VNO, whose customer relationship management system records the history of technician interventions at the customer's premises as in Table 5.1. For each problem, the following attributes are available at least.

Table 5.1   Log of technicians' on-site interventions (techdb).

Attribute	Description
Date	Opening date of the intervention
PBX-ID	Unique ID of the private branch exchange
CType	Customer's line of business
Tech-ID	Unique ID of technician's interventions
Severity	Problem severity recorded after problem solution
Prob-ID	Problem type recorded after problem solution

Table 5.2   Values of the severity field in Table 5.1.

Severity	Meaning
1	Low-level problem
2	Medium, intermittent service interruptions
3	Critical, service unavailable

Attribute Date consists of the problem opening date and time, defined as the time the call center receives a customer call reporting the malfunction. PBX-ID represents a unique identifier of the PBX involved. If more than one PBX has been installed by a customer, this is determined by the call-center operator based on the customer's description of the problem and the available configuration of PBXs installed at the customer's premises.

CType is the customer's line of business, including banks and health care, insurance and telecommunication businesses. Attribute Tech-ID is a unique identifier of technician interventions: during the same intervention one or more problems may be tackled. The Severity value is a measure of problem impact. It is defined on a scale from 1 to 3, as reported in Table 5.2. Finally, Prob-ID is a coding of the malfunction solved by the technician; 200 problem descriptions are codified. It is worth noticing that the closing date of the intervention is missing.

5.3.2.2   MAP

Since 200 problem types may be too fine-grained details for OpR analysis, problem types are organized into a three-level hierarchy, as shown in Table 5.3. The lowest level is the problem type reported without any modification by the technician database. The highest level (EC1) consists of the Basel II event categories for IT operational risk: *software, hardware, interface, security* and *network communications*. The middle level (EC2) is an intermediate categorization level.

Every problem type readily falls into one of the five EC1 level members. However, the mapping tool also has to accept the specification of a previously unseen problem type, to be stored and recorded in the data merging repository.

Table 5.3   Correspondences between the problem description and
EC1 and EC2 categories (map).

Attribute	Description
Problem	Problem specification
EC1	High-level EC1 category of the specified problem
EC2	Middle-level EC2 category of the specified problem

Table 5.4   Log of PBX alarms (alarm).

Attribute	Description
PBX-ID	Unique ID of the private branch exchange
TestDate	Date and time a log was downloaded
Alarms	Sequence of alarms raised since last log download

5.3.2.3 ALARM

The third data source is the collection of log files generated by PBXs. The files are
downloaded into a centralized repository, called alarm, on a regular round-robin
basis (see Table 5.4).

For a given PBX identifier PBX-ID and date–time TestDate when a log is
downloaded, alarm stores the set of alarms (Alarms) raised by the PBX since
the previous log download. Sixteen distinct alarms are available in the data.

Unfortunately, the precise time at which an alarm is raised is not stored in
the PBX log.

5.3.2.4 Balance sheet information

The last data source includes financial indicators derived from the balance sheets
of small and medium-sized enterprises, interested in new services or in renewing
contracts. Financial balance sheets as input for the integration are supposed to
be in the machine-processable XBRL standard (XBRL, 2010). Moreover, since
only a subset of the information encoded in the XBRL document is relevant
as an output of the merging, we have to restrict our integration to a limited
number of financial indicators. In other words, which attributes to select is an
outcome of the development of the statistical model and/or a parameter of the
data merging service.

Table 5.5 summarizes an example of the financial indicators available for a
given customer of the service, determined by the PBX-ID attribute.

5.3.3   Data merging for risk assessment

Assessing the risk of business in small and medium-sized enterprises requires,
first of all, integrating data from heterogeneous sources. In our context, data

Table 5.5   Balance sheet indicators of a given customer of the VNO (balance).

Attribute	Description
PBX-ID	Unique ID of the private branch exchange
ReturnOnEquity	Measures the rate of return on the ownership interest
NonCurrentAssess	An asset which is not easily convertible to cash or not expected to become cash within the next year
NetCashFromRegularOperations	Cash flow available for debt service – the payment of interest and principal on loans
NetProfit	The gross profit minus overheads minus interest payable plus/minus one-off items for a given time period
CurrentRatio	Measures whether or not a firm has enough resources to pay its debts over the next 12 months
CurrentAssets	An asset on the balance sheet that is expected to be sold or otherwise used up in the near future
CurrentLiabilities	Liabilities of the business that are to be settled in cash within the fiscal year or the operating cycle
Pre_Tax_Profit	The amount of profit a company makes before taxes are deducted

includes the customer type, the amount of service calls, the related malfunctions and, as a final dimension, the financial strength of such a customer. The issue of merging and integrating these various sources is critical for guaranteeing the quality of the final data that will be the subject of the risk analysis.

The flow of data in the merging activities is shown in Figure 5.2. According to the service-oriented architecture of the MUSING platform, the software is implemented as a web service. In this way, the inputs are required to be provided as XML documents compliant with a public XML schema, rather than in a proprietary format (e.g. text files and Excel sheets).

Three main activities can be recognized in the database merging execution:

1. <MAP MERGER> integrates the problem type hierarchy stored in map with techdb. The output of such an integration is a new database, named annotated, containing the same fields as techdb plus the EC1 and EC2 attributes related to the problem specification.

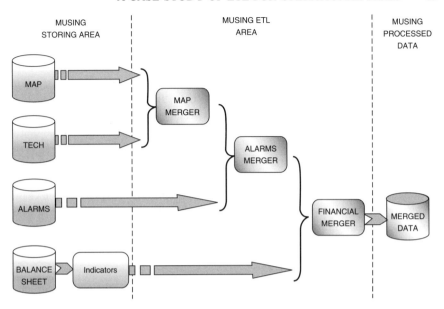

Figure 5.2   The three-step process of the database merging activity.

2. <ALARMS MERGER> uses the output generated by the previous step to produce an integrated database called service-call. The latter adds the sequences of the alarms generated by the PBX loaded from the alarm database to the logs of technicians' on-site interventions.

3. Finally, service-call is integrated with the financial indicators stored in the balance database. This step is concretely achieved by means of the <FINANCIAL MERGER> module using a MUSING component (<INDICATORS> in Figure 5.2) to parse and select the required indicators from the XBRL balance sheets. The output of such a component is a new XML document storing the required financial information.

5.3.4   The issues of data merging in MUSING

The tasks described in the previous section create several issues related to the merging activity, principally due to the information stored in the available data sources.

The first step does not introduce any particular problem, since the join between the map database and the technician database is executed considering the PROBLEM and PBX-ID fields, respectively. On the contrary, let us consider the second step. From a database perspective, only PBX-ID can be seen as a primary key for the available data. Unfortunately, for the <ALARM MERGER> component, this information is not enough to guarantee correct integration, since there is no external key to bind an intervention (performed by a technician) exactly to a sequence of alarms.

In other words, the main question is how to merge the alarms considering the opening date contained in the technician's database. Unfortunately, while an intervention records the timestamp when it occurred, the alarm log contains the timestamp when an alarm is downloaded to the central repository, not the timestamp when the alarm is raised. More correctly, the alarm log collection may occur once every few days, and only the alarms raised from one download to another are available for each entry of Table 5.4.

Moreover, even considering the date–time (available in Table 5.4) as the time when the alarms actually occur, we cannot compute a precise merging, since the important information related to when the technician intervention has been closed is missing in the technician database. This makes it impossible to know exactly which temporal window must be considered to check the sequences of alarms in the time interval under analysis.

The above observations can be synthesized as:

1. *Issues.* (1) The lack of information to accurately bind the technician intervention with the sequences of alarms and (2) the lack of the exact time when the alarm is raised.

2. *Solution.* Due to these problems, a heuristic is necessary to enable the merging of the alarms. As represented in Figure 5.3, two parameters, namely *DInf* and *DSup*, have been introduced to enable the system to compute the temporal window for executing the analysis and then merging the alarm sequences. In particular, given the problem opening date value t stored in the technician intervention, a temporal window is computed, adding the number of days defined by the *DSup* parameter to the available date, and then subtracting *Dinf* days from the same value. These two parameters are necessary, since we want to discover the sequence of alarms that cause the intervention, and then from the opening date we have to consider a period before the problem is noticed. *DSup* is necessary also to consider the alarms that occur during a period after the intervention request is opened.

By introducing this heuristic, the system is now able to merge the alarms that have been downloaded by the PBX monitoring system on a date included in the time window considered.

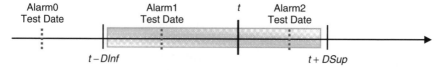

Figure 5.3   Database merging temporal window.

Example: Let us consider the following intervention:

```
[Date:12/02/2009, PBX-ID:90333, CType:Bank, Severity:2,
                 Prob-ID:Hardware]
```

and the following sequences of alarms downloaded by the system monitoring device for the PBX under analysis:

```
90333, 19/02/2007, {DIGITAL TRUNK CARD,DKT SUBUNIT}
90333, 23/02/2007, {POWER SUPPLY}
```

If we consider *DInf* = *30* and *DSup* = *10* days, given the date value in the intervention, all the sequences of alarms raised for PBX 90333 downloaded on a date included in the [12/01/2009-22/02/2009] interval are added to the previous intervention, and only the sequence {DIGITAL TRUNK CARD,DKT SUBUNIT} is added to the specific intervention entry.

It is worth noting that, by varying these two parameters, the analyst can consider different hypotheses and can observe whether there is a sequence of alarms always related to the specific problem by taking into account different time windows. The lack of a 'closing date' in Table 5.1 (which would have completely defined the temporal window for analysis in its entirety) introduces a sort of approximation between the real alarms raised and the sequence of alarms available in the merged database. As mentioned above, the analyst can nevertheless discover the correlation between the alarms and the problems reported in Table 5.1.

5.4   The ETL-based solution

The solution proposed in this case study employs an ETL job (shown in Figure 5.4), related to the main steps of the data merging. The job design reflects the needs of OpR and essentially respects the flow reported in Figure 5.2.

This job starts with a first transformation, named problemsAnnotation, for annotating the technician database by adding the categories EC1 and EC2, available in Table 5.3, for each known problem.

Subsequently, the merging of the technician database with the alarms database works by means of the alarmsAppender transformation. Such a merging is based on PBX-ID and the date when the technician intervention has been opened. As stated in the previous section, the closing date is not available for any specific intervention, so a heuristic is employed to extract and merge the sequences of alarms for a specific PBX.

Finally, the last step involves the merging of the financial indicators with the PBXs and their subsequent serialization (financialAppender).

It must be pointed out that the cited job also manages the error handling (e.g. XML input validation) and the transformations can accept some arguments as

Figure 5.4   The `dataMerging` *job.*

parameters. Moreover, different jobs are defined to execute only a part of the entire process.

The available jobs are executed by invoking a web service that sends the specific tasks (i.e. transformations) to a Pentaho Job Executor Service. This modularity is quite powerful, since it enables the user to modify only the Pentaho transformations when a change is required, for example due to a variation of a datatype.

In the following sections, the three Pentaho transformations employed for data merging are outlined.

5.4.1   Implementing the 'map merger' activity

The transformation that adds the `EC1` and `EC2` categories to the problem descriptions is shown in Figure 5.5. It takes the list of problems from Table 5.3 and merges the available ones with the problems defined in Table 5.1. Two simple XML source fragments are shown in Figure 5.6. The information required is loaded via simple XPath expressions used to locate both the input rows and the fields inside each row. The specification of such expressions is encoded in the map and the `techdb` operators. For example, the reported XPath expression

```
doc("map.xml")/Rows/Row/EC1/text()
```

denotes the string values contained in the tag `<EC1>` of map (see also Figure 5.6 (right)).

The problem description field is the key value for the subsequent merging (`mergeProblems` operator). Since it is not guaranteed that all the available problems in Table 5.1 have a correspondence in Table 5.3, the values of `EC1` and `EC2`

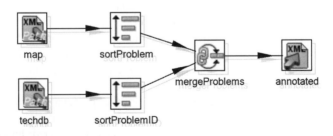

Figure 5.5   The `problemsAnnotation` *transformation.*

```
<Rows>
  <Row>
    <PBX_No>91717</PBX_No>
    <Opening_Date>18/02/2007 09:22</Opening_Date>
    <Severity>3</Severity>
    .....
    <Problem> Administrator problem </Problem>
  </Row>
  <Row>
    <PBX_No>91717</PBX_No>
    <Opening_Date>19/02/2007 09:29</Opening_Date>
    <Severity>1</Severity>
    ....
    <Problem>Handling replacement equipment</Problem>
  </Row>
  .....
</Rows>
```

```
<Rows>
  <Row>
    <Problem>ACD display board malfunctioning</Problem>
    <EC2>INT</EC2>
    <EC1>Interface</EC1>
  </Row>
  <Row>
    <Problem>Administrator problem</Problem>
    <EC2>SEC</EC2>
    <EC1>Security</EC1>
  </Row>
  .....
</Rows>
```

Figure 5.6 XML fragments of the technician database (left) and the map database (right).

for missing problems are set to `null`. The `sortProblem` and `sortProblemId` operators introduced in the transformation are necessary to sort the fields used as keys before computing the join.

The result of this transformation is an XML file that adds the categories `EC1` and `EC2` defined in Table 5.3 to each problem originally available in the technician database. This file is serialized by means of the `annotated` operator, in which we encode the required output via XPath expressions, defined over the stream of data produced by the `mergeProblems` operator.

5.4.2 Implementing the 'alarms merger' activity

Figure 5.7 shows the merging activity between the annotated file produced at the previous step (`annotated`) and the logs generated by the PBX (`alarm`). The `getInfSup` operator loads the parameters defined in the heuristic, in order to consider the time interval to check the alarm sequences (see Section 5.3.4).

In particular, given two integers, namely *DInf* and *DSup*, the transformation adds two additional fields recording the starting and ending date of the alarm monitoring (`computeDates`) into the result of the `annotated` operator. These values are computed by adding (resp. subtracting) the *DInf* (resp. *DSup*) value to the opening date defined in the annotated database.

Once the date bounds are computed, alarms are merged with the technician interventions by considering the specific PBX identifier and the date when the alarms have been downloaded by the monitoring system. If the date is included in the input bounds of the analysis, the sequence of alarms is added to the intervention.

The output generated is stored to be available for the next step of the job, in which some financial indicators extracted from the balance sheets are added to conclude the data merging for integrated operational risk management.

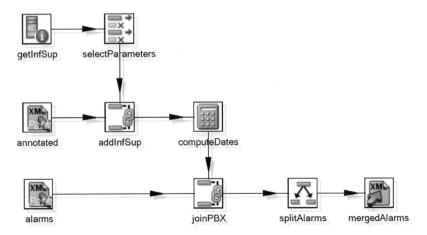

Figure 5.7   The `alarmsAppender` *transformation.*

5.4.3   Implementing the 'financial merger' activity

The transformation depicted in Figure 5.8 receives as input three different XML sources: (1) the merged file produced by the `alarmsAppender` transformation, including the sequences of alarms (`mergedAlarms`); (2) the set of indicators extracted from the balance sheets (`indicators`); and finally (3) some qualitative scores computed and provided via an external service (`scores`). It returns the final result of the merging activities, as represented in Figure 5.9. Starting from the initial data, we add several fields to enhance the OpR data sets. The final output is a database in which it is possible to discover the causes of a problem, analyzing the alarms raised by the specific problem and its impact on the enterprise activities. The financial indicators enable the analyst to study different scenarios for designing specific approaches to mitigate the risk related to IT, based also on the particular business features of the enterprise under analysis.

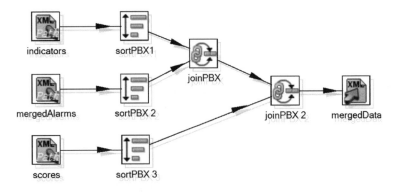

Figure 5.8   The `financialAppender` *transformation.*

PBX - Id	Opening Date	⋯	Problem
90210	10-feb-2009		Bad card
91205	05-mar-2009		No audio

problemAnnotation

PBX - Id	Opening Date	⋯	Problem	EC1	EC2
90210	10-feb-2009		Bad card	Hardware	HD01
91205	05-mar-2009		No audio	Interface	INT04

alarmsAppender

PBX - Id	Opening Date	⋯	Problem	EC1	EC2	Alarms
90210	10-feb-2009		Bad card	Hardware	HD01	$a_i, a_j, \ldots, a_k$
91205	05-mar-2009		No audio	Interface	INT04	$a_x, a_y, \ldots, a_z$

financialAppender

PBX-Id	Opening Date	⋯	Problem	EC1	EC2	Alarms	Indicators
90210	10-feb-2009		Bad card	Hardware	HD01	$a_i, a_j, \ldots, a_k$	$i_1, i_2, \ldots, i_n$
91205	5-mar-2009		No audio	Interface	INT04	$a_x, a_y, \ldots, a_z$	$i_1, i_2, \ldots, i_n$

Figure 5.9   The data merging output.

5.5 Summary

In this chapter we introduce an application of typical ETL processes to perform operational risk analysis by means of a case study in the context of a VNO. The idea is to merge the data available from heterogeneous and different sources to enable the design of OpR schemas, with special reference to IT management. In our discussion, the case study is outlined from two different perspectives. On the one hand, we focus on the logical structure of the data, on the information available and on the high-level architecture. On the other hand, we propose several details of the implemented solution, describing the main transformations of data merging and the results obtained.

From the OpR perspective, our application represents a powerful tool for analysts and can be used for merging financial data to enable a differentiation of the strategies adoptable, while considering several financial aspects related to an enterprise. The proposed solution is modular, since it enables the user to change only the Pentaho transformations, and not the complete system, when a change is required. An example of the usage of this application can be found in Grossi, Romei and Ruggieri (2008), in which the merging tool is used as a base to prepare data for sequential pattern mining.

References

Albrecht, A. and Naumann, F. (2008) Managing ETL Processes, in *Proceedings of the International Conference on Very Large Data Bases, VLDB 2008*, Auckland, New Zealand, pp. 12–15.

Azvine, B., Cui, Z., Majeed, B. and Spott, M. (2007) Operational Risk Management with Real-Time Business Intelligence, *BT Technology Journal*, 25, 1, pp. 154–167.

Basel Committee on Banking Supervision (2006) *International Convergence of Capital Measurement and Capital Standards: A Revised Framework*, www.bis.org/publ/bcbs128.pdf?noframes=1 (accessed 6 March 2010).

Bernstein, P.A. (2001) Generic Model Management: A Database Infrastructure for Schema Manipulation, in *Proceedings of the International Conference on Cooperative Information Systems, CoopIS 2001*, Trento, Italy, pp. 1–6.

Dessloch, S., Hernández, M.A., Wisnesky, R., Radwan, A. and Zhou, J. (2008) Orchid: Integrating Schema Mapping and ETL, in *Proceedings of the International Conference on Data Engineering, ICDE 2008*, Cancún, Mexico, pp. 1307–1316.

Grossi, V., Romei, R. and Ruggieri, R. (2008) A Case Study in Sequential Pattern Mining for IT-Operational Risk, in *Proceedings of the European Conference on Machine Learning and Principles and Practice of Knowledge Discovery in Databases, ECML/PKDD 2008*, Antwerp, Belgium, pp. 424–439.

IBM WebSphere DataStage (2010) www.ibm.com/software/data/infosphere/datastage (accessed 21 May 2010).

Informatica PowerCenter (2010) http://www.informatica.com/powercenter (accessed 6 March 2010).

Kimball, R. and Caserta, J. (2004) *The Data Warehouse ETL Toolkit*, John Wiley & Sons, Ltd, Chichester.

Kof, L. (2005) Natural Language Processing: Mature Enough for Requirements Documents Analysis?, in *Proceedings of the International Conference on Applications of Natural Language to Information Systems, NLDB 2005*, Alicante, Spain, pp. 91–102.

Luján-Mora, S., Vassiliadis, P. and Trujillo, J. (2004) Data Mapping Diagrams for Data Warehouse Design with UML, in *Proceeding of the International Conference on Conceptual Modeling, ER 2004*, Shanghai, China, pp. 191–204.

Melnik, S. (2004) *Generic Model Management: Concepts and Algorithms*, Springer-Verlag, Berlin.

MUSING (2006) IST-FP6 27097, http://www.musing.eu (accessed 21 May 2010).

Oracle Warehouse Builder (2010) http://otn.oracle.com/products/warehouse/content.html (accessed 21 May 2010).

Pentaho (2010) Open Source Business Intelligence, www.pentaho.com (accessed 11 January 2010).

Sellis, T.K. and Simitsis, A. (2007) ETL Workflows: From Formal Specification to Optimization, in *Proceedings of the East-European Conference on Advances in Databases and Information Systems, ADBIS 2007*, Varna, Bulgaria, pp. 1–11.

Simitsis, A., Vassiliadis, P. and Sellis, T.K. (2005) Optimizing ETL Processes in Data Warehouses, in *Proceedings of the International Conference on Data Engineering, ICDE 2005*, Tokyo, Japan, pp. 564–575.

Simitsis, A., Vassiliadis, P., Skiadopoulos, S. and Sellis, T.K. (2006) *Data Warehouses ad OLAP: Concepts, Architectures and Solutions*, Data Warehousing Refreshment, IRM Press, Hershey, PA.

Simitsis, A., Skoutas, D. and Castellanos, M. (2008) Natural Language Reporting for ETL Processes, in *Proceedings of the International Workshop on Data warehousing and OLAP, DOLAP'08*, Napa Valley, California, pp. 65–72.

Vassiliadis, P., Simitsis, A. and Skiadopoulos, S. (2002) Conceptual Modeling for ETL Processes, in *Proceedings of the International Workshop on Data warehousing and OLAP, DOLAP 2002*, McLean, Virginia, pp. 21–28.

Wisnesky, R., Radwan, A., Hernandez, M.A., Dessloch, S. and Zhou, J. (2008) Orchid: Integrating Schema Mapping and ETL, in *Proceedings of the International Conference on Data Engineering, ICDE 2008*, Cancún, Mexico, pp. 1307–1316.

XBRL (2010) eXtensible Business Reporting Languages, http://www.xbrl.org (accessed 6 March 2010).

6

Risk-based testing of web services

Xiaoying Bai and Ron S. Kenett

6.1 Introduction

Service-oriented architecture (SOA), and its implementation in web services (WS), introduce an open architecture for integrating heterogeneous software through standard Internet protocols (Papazoglou and Georgakopoulos, 2003; Singh and Huhns, 2005). From the providers' perspective, proprietary in-house components are encapsulated into standard programmable interfaces and delivered as reusable services for public access and invocation. From the consumers' perspective, applications are built following a model-driven approach where business processes are translated into control flows and data flows, of which the constituent functions can be automatically bound to existing services discovered by service brokers. In this way, large-scale software reuse of Internet-available resources is enabled, providing support for agile and fast response to dynamic changing business requirements. SOA is believed to be the current major trend of software paradigm shift.

Due to the potential instability, unreliability and unpredictability of the open environment in SOA and WS, these developments present challenging quality issues compared with traditional software.

Traditional software is built and maintained within a trusted organization. In contrast, service-based software is characterized by dynamic discovery and composition of loosely coupled services that are published by independent providers.

Operational Risk Management: A Practical Approach to Intelligent Data Analysis Edited by Ron S. Kenett and Yossi Raanan © 2011 John Wiley & Sons, Ltd

A system can be constructed, on-the-fly, by integrating reusable services through standard protocols (O'Reilly, 2005). For example, a housing map application can be the integration of two independent services: Google Map service (Google, 2010) and housing rental services (Rent API, 2010). In many cases, the constituent data and functional services of a composite application are out of the control of the application builder. As a consequence, service-based software has a higher probability to fail compared with in-house developed software. On the other hand, as services are open to all Internet users, the provider may not envision all the usage scenarios and track the usage status at runtime. Hence, a failure in the service may affect a wide range of consumers and result in unpredictable consequences. For example, Gmail reported a failure of 'service unavailable due to outage in contacts system' on 8 November 2008 for 1.5 hours – millions of customers were affected.

Testing is thus important to ensure the functionality and quality of the individual services as well as composed integrated services. Proper testing can ensure that the selected services can best satisfy the users' needs and that services that are dynamically composed can interoperate with each other. However, testing is usually expensive and confidence in a specific service is hard to achieve, especially in an open Internet environment. The users may have multiple dimensions of expected features, properties and functional points that result in a large number of test cases. It is both time and resource consuming to test a large set of test cases on a large number of service candidates.

To overcome these issues, the concept of group testing was introduced to services testing (Bai *et al.*, 2007a, 2007b; Tsai *et al.*, 2004, 2005). With this approach, test cases are categorized into groups and activated in groups. In each group testing stage, the failed services are eliminated through a predefined ruling-out strategy. Effective test case ranking and service ruling-out strategies remove a large number of unreliable services at the early stage of testing, and thus reduce the total number of executed tests. The key problem in progressive group testing is the ranking and selection of test cases.

Measurement is essential for test ranking and selection. Statistical research provides promising techniques for qualitative measuring and ranking of test cases and services. Kenett and Zacks (1998) and Kenett and Steinberg (2006) present the applications of statistics and DoE (Design of Experiments) methods in modern industry in general, and in software development in particular (Kenett and Baker, 2010). In particular, Bayesian techniques such as Bayesian inference and Bayesian networks can help in estimation and prediction. Harel *et al.* (2008) apply such procedures in the context of web usability assessment. They further refine the controlled process with usability change tracking, service indices, usability diagnosis and corrective adjustments (Kenett *et al.*, 2009).

This chapter describes a risk-based approach to evaluate services quantitatively and rank test cases. Risk-based testing was proposed to select and schedule test cases based on software risk analysis (Amland, 2000; Bach, 1999; Chen and Probert, 2003; Ottevanger, 1999; Redmill, 2004; Rosenberg *et al.*, 1999). With this approach, test cases are prioritized and scheduled based on the risks of their target software features. When time and resources are limited, test engineers can select a subset of test cases with the highest risk targets, in order to achieve good enough quality with affordable testing effort. This approach is similar to the study of risks of rare events ('black swans') in the financial industry (Taleb, 2007, 2008) and risk management in general as presented in Chapters 1, 11 and 14. Kenett and Tapiero (2010) propose a convergence between risk engineering and quality control from the statistical perspective, which is also discussed in Chapter 14. In software engineering, risk-based software testing has been gaining attention since the late 1990s. Amland (2000) established a generic risk-based testing approach based on Karolak's risk management process model (Karolak, 1996). Bach (1999) identifies the general categories of risks during software system development including complexity, change, dependency, distribution, third party, etc.

This chapter is about the application of risk-based techniques to testing WS. Figure 6.1 presents an overview of the approach. As shown in the figure, the service-based system is specified from three perspectives: the interface model of the exposed functionalities of individual or composite service, the workflow model of service composition logic, and the semantic model to define the concepts and domain knowledge for service mutual understanding and interoperation. In particular, the chapter addresses the problem of risk-based test ranking and selection in the context of semantic WS. Risks are assessed based on the semantic model of service data, interface operation and workflow. The risks of the services are identified in two ways, static and dynamic analysis. Static analysis is based on service dependencies and usage topologies. As services may recomposed online, dynamic analysis is introduced to detect the changes at runtime and recalculate the service risks based on runtime profiling of service composition, configuration and operation. Test cases are associated to their target features under test and are ranked based on the risks of the services.

An important issue in the risk-based approach is the measurement of risk. In most cases, the risk factors are estimated subjectively based on experts' experiences. Hence, different people may produce different results and the quality

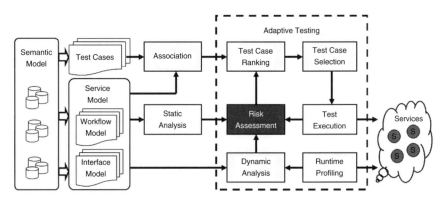

Figure 6.1   Approach overview.

of risk assessment is hard to control. Some researchers began to practise the application of ontology modelling and reasoning to operational and financial risk management, combining statistical modelling of qualitative and quantitative information on an SOA platform (Spies, 2009) and the MUSING project (MUSING, 2006). In this chapter, we present an objective risk measurement based on ontology and service analysis. A Bayesian network (BN) is constructed to model the complex relationships between ontology classes. Failure probability and importance are estimated at three levels – ontology, service and service workflow – based on their dependency and usage relationships.

In addition to the traditional issues, the open and dynamic nature of WS imposes new challenges to software testing techniques. In the open platform, the service consumer cannot fully control services that are remotely managed and maintained by some unknown providers. For example, in early research on software reliability, Kenett and Pollak (1986) discovered that software reliability decays over time due to the side effects of bug correction and software evolution. In a service-based system, such decaying process may not be visible to the consumers until a failure occurs. Services are changed continuously online (the 'The Perpetual Beta' principle in Web 2.0, O'Reilly, 2005). The composition and collaboration of services are also built on demand. As a result, testing could not be fully acknowledged beforehand and has to be adaptive so that tests are selected and composed as the services change. The chapter introduces the adaptive testing framework into the overall process so that it can monitor the changes in service artefacts and dependencies, and reactively reassess their risks and reschedule the test cases for group testing.

More and more researchers are beginning to realize the unique requirements of WS testing and propose innovative methods and techniques from various perspectives, such as collaborative testing architecture, test case generation, distributed test execution, test confidence and model checking (Canfora and Penta, 2009). Most of current WS testing research focuses on the application and adaptation of traditional testing techniques. We address here the uniqueness of WS and discuss WS testing from a system engineering and interdisciplinary research perspective. This chapter is part of a more general convergence between quality engineering and risk methodologies. Compared with existing work in this area, the chapter covers the following aspects:

- It proposes an objective method to assess software risks quantitatively based on both the static structure and dynamic behaviour of service-based systems.

- It analyses the quality issues of semantic WS and measures the risks of the ontology-based software services based on semantic analysis using stochastic models like Bayesian networks.

- It improves WS progressive testing with effective test ranking and selection techniques.

The chapter is organized as follows. Section 6.2 introduces the research background of risk-based testing, progressive group testing and semantic WS. Section 6.3 discusses the problem of adaptive WS testing. Section 6.4 analyses the characteristics of semantic WS and proposes the method for estimating and predicating failure probability and importance. Section 6.5 discusses the method of adaptive measurement and adaptation rules. These techniques are the basis for incorporating a dynamic mechanism into the WS group testing schema. Section 6.6 presents the metrics and the experiments to evaluate the proposed approach. Section 6.7 concludes the chapter with a summary and discussion.

6.2 Background

This chapter is about risk-based testing, service progressive group testing and semantic WS. We begin with an introduction of these important topics of application of operational risk management.

6.2.1 Risk-based testing

Testing is expensive, especially for today's software with its growing size and complexity. Exhaustive testing is not feasible due to the limitations of time and resources. A key test strategy is to improve test efficiency by selecting and planning a subset of tests with a high probability of finding defects. However, selective testing usually faces difficulties in answering questions like 'What should we test first given our limited resources?' and 'When can we stop testing?' (Kenett and Pollak, 1986; Kenett and Baker, 2010).

Software risk assessment identifies the most demanding and important aspect of the software under test, and provides the basis for test selection and prioritization. In general, the risk of a software feature is defined by two factors: the probability to fail and the consequence of the failure. That is,

$$Risk(f) = P(f) * C(f) \tag{6.1}$$

where f is the software feature, $Risk(f)$ is its risk exposure, $P(f)$ is the failure probability and $C(f)$ is the cost of the failure. Intuitively, a software feature is risky if it has a high probability to fail or its failure will result in serious consequences. Intuitively, a risky feature deserves more testing effort and has a high priority to test. Risk-based testing was proposed to select and schedule test cases based on software risk analysis (Amland, 2000; Bach, 1999; Chen and Probert, 2003; Ottevanger, 1999; Redmill, 2004; Rosenberg et al., 1999; Kenett and Baker, 2010). The general process of risk-based testing is as follows:

1. Identify the risk indicators.

2. Evaluate and measure the failure probability of software features.

3. Evaluate and measure the failure consequence of software features.

4. Associate the test cases to their target software features.

5. Rank the test cases based on the risks of their target features. Test cases with risky features should be exercised earlier.

6. Risk-related coverage can be defined to control the testing process and test exit criteria.

Practices and case studies show that testing can benefit from the risk-based approach in two ways:

• The reduced resource consumption.

• Improved quality by spending more time on critical functions.

6.2.2   Web services progressive group testing

The group testing technique was originally developed at Bell Laboratories for efficiently inspecting products (Sobel and Groll, 1959). The approach was further expanded to general cases (Dorfman, 1964; Watson, 1961). It is routinely applied in testing large number of blood samples to speed up the test and reduce the cost (Finucan, 1964). In that case, a negative test result of the group under test indicates that all the individuals in the group do not have the disease; otherwise, at least one of them is affected. Group testing has been used in many areas such as medical, chemical and electrical testing, coding, etc., using either combinatorial or probabilistic mathematical models.

Tsai et al. (2004) introduce the ideas of progressive group testing to WS testing. We define test potency as its capability to find bugs or defects. Test cases are ranked and organized hierarchically according to their potency to detect defects, from low potency to high. Test cases are exercised layer by layer, following a hierarchical structure, at groups of services. Ruling-out strategies are defined so that WS that fail at one layer cannot enter the next testing layer. Test cases with high potency are exercised first with the purpose to remove as many WS as early as possible. Group testing is by nature a selective testing strategy, which is beneficial in terms of a reduced number of test runs and shortened test time.

WS progressive group testing enables heuristic-based selective testing in an open service environment (Tsai et al., 2003a, 2003b, 2004; Bai et al., 2007b). To refine the model, Bai et al. (2007b) discuss the ranking and selecting strategy based on test case dependency analysis. Two test cases $tc1$ and $tc2$ are dependent if a service that fails $tc1$ will also fail $tc2$. Hence, for any service, a test case will be exercised only if the service passes all its dependent test cases. They further propose the windowing technique that organizes WS in windows (Tsai et al., 2005) and incorporates an adaptive mechanism (Bai et al., 2007b; Bai and Kenett, 2009) which follows software cybernetics theories (Cai, 2002; Cangussu *et al.*, 2009), in order to adjust dynamically the window size, determine WS ranking and derive test case ranking. Several challenges need to be addressed in setting up progressive group testing strategies, including:

1. Estimation of probabilities.

2. Specification of dependencies.

3. Dynamic updating of estimates.

4. Sensitivity evaluation of group testing rule parameters.

6.2.3   Semantic web services

The Semantic Web is a new form of web content in which the semantics of information and services are defined and understandable by computers, see Chapter 4 and Berners-Lee *et al.* (2008). Ontology techniques are widely used to provide a unified conceptual model of web semantics, see Chapter 3, Gomez-Perez *et al.* (2005) and Spies (2009). In WS, OWL-S provides a semantic model for composite services based on the OWL (Web Ontology Language) specification. OWL-S specifies the intended system behaviour in terms of inputs, outputs, process, pre-/post-conditions and constraints using the ontology specifications of data and services (Martin, 2004; W3C, 2010a, 2010b). Spies (2009) and the MUSING project (MUSING, 2006) introduce a comprehensive approach of ontology engineering to financial and operational risks as described in Chapter 3 and Chapter 4.

However, semantics introduce additional risks to WS. First, the ontologies may be defined, used and maintained by different parties. A service may define the inputs/outputs of its interface functions as instances of ontology classes in a domain model that is out of the control of the service provider and consumer. Due to the complexity of conceptual uniformity, it is hard to ensure completeness and consistency, and the unified quality of the ontologies. For example, AAWS (Amazon Associate Web Services) is an open service platform for online shopping. Its WSDL service interface provides 19 operations with complicated data structure definitions. In this research, we translated the WSDL data definitions to ontology definitions for semantic-based service analysis. We identified 514 classes (including ontology and property classes) and 1096 dependency relationships between the classes.

Moreover, ontologies introduce complex relationships among the data and services which result in the increased possibility of misuse of the ontology classes. For example, in the domain, two ontology classes 'CourseBook' and 'EditBook' are defined with inheritance from the 'Book' ontology. From the domain modeller perspective, the two categories of books are used for different purposes and are mutually exclusive. However, from the user perspective, such as a courseware application builder, the two classes could be overlapped because an edited book can also be used as reading material for graduate students. Such conflicting views will result in a possible misuse of the class when developing education software systems.

6.3 Problem statement

Given a set of services $S = \{s_i\}$ and a set of test cases $T = \{t_i\}$, selective testing is the process of finding an ordered set of test cases to detect bugs as early as possible, and as many as possible. However, testing and bug detection are like a 'moving target' shooting problem. As testing progresses, bugs are continuously detected and removed. As a result, the bug distribution and service quality are changed. On the other hand, the bug detection potency of each test case is also changed, as each test case may be effective in identifying different types of bugs. Adaptive testing is the mechanism to calculate the ranking and ordering of test cases, at runtime, so as to reflect the dynamic changes in the services, test cases potencies and bugs.

Suppose that $\forall s \in S$, $B(s) = \{b_i\}$ is the set of bugs in the service, $T(s) \subseteq T$ is the set of test cases for the service s, $\forall b \in B(s)$, $\exists T(b) \subset T(s)$ so that $\forall t_i \in T(b)$, t_i can detect b. Ideally, the potency of a test case t, $\wp(t)$, is defined as the capability of a test case to detect bugs. That is,

$$\wp(t) = \frac{|B(t)|}{\sum |B(s_i)|} \qquad (6.2)$$

where $B(t)$ is the set of bugs that the test case t can detect, $|B(t)|$ is the number of bugs t can detect, and $\sum |B(s_i)|$ is the total number of bugs in the system. The problem is to find an ordered subset of T so that $\forall t_i \in T$ and $t_j \in T$, $0 \le i, j \le n$, if $i < j$ then $\wp(t_i) > \wp(t_j)$.

However, it is usually hard to obtain an accurate number of bugs present in the software that the test case can detect. In this work, we transform the bug distribution problem to a failure probability so that, rather than measuring the number of bugs in a service, we measure the failure probability of the service.

We further consider the failures' impact and combine the two factors into a risk indicator for ranking services. In this way, testing is a risk mitigation process. Suppose that $Risk(s) = P(s) * C(s)$ is the risk of a service, so the potency of a test case is defined as

$$\wp(t) = \frac{\sum Risk(ts_i)}{\sum Risk(s_i)} \qquad (6.3)$$

where ts_i is the set of services that t tests. To define the process, we make the following simplified assumptions:

1. A service can have at most n bugs.

2. The bugs are independent.

3. A bug is removed immediately after being detected.

The general process of risk-based test selection is as follows:

1. Calculate the risks of each service, and order the services in a sequence s_i such that, $\forall s_i, s_j \in S, 0 \le i, j \le n$, if $i < j$ then $Risk(s_i) > Risk(s_j)$.

2. Select the sets of test cases for each service and order the test sets in sequence T_i such that $T_i = T(s_i)$ and $\forall T_i, T_j \subseteq T, 0 \le i, j \le n$, if $i < j$ then $Risk(s_i) > Risk(s_j)$.

3. Rank the test cases in each test set according to their potencies. That is, $T_i = \{t_{ij}\}$ such that $\forall t_i j, t_i k \in T_i, 0 \le j, k \le n$, if $j < k$ then $\wp(t_{ij}) > \wp(t_{ik})$).

4. Select the service s_i with the highest risk and select the corresponding set of test cases T_i. Exercise the test cases in T_i, in sequence.

5. Recalculate the service risks and test case potencies.

6. Repeat steps 4–5 until certain criteria are met. We can define different exit criteria such as the percentage of services covered, the percentage of test cases exercised, the number of bugs detected, etc.

6.4 Risk assessment

Based on the analysis of semantic services, risks are assessed at three layers (data, service and composite service) from two perspectives (failure probability and importance).

6.4.1 Semantic web services analysis

A failure in service-based software may result from many factors such as misused data, unsuccessful service binding and unexpected usage scenarios. To calculate the risk of services-based software, this section analyses the features of semantic WS from the following three layers:

1. The domain layer: that is, how ontologies are defined and dependent on each other.

2. The service layer: that is, how ontologies are instantiated and used as the input/output data to each services.

3. The composition layer: that is, how services are organized and affect each other in a control structure in a workflow.

6.4.1.1 Ontology dependency analysis

Ontology techniques are widely used to specify the conceptual model and semantic restrictions of a service domain (see Chapter 3). A domain usually contains a large number of ontologies with complex relationships. Such a relationship specification is usually a manual labelling process. An error in a class definition may affect others and propagate to a large scale along the dependencies (see Chapter 9 on association rules and Chapter 10 on near miss models). To analyse the robustness of the model, we need to understand how ontologies relate to each other. In general, the dependency relationships are identified from the following perspectives:

- **Inheritance.** OWL and RDFS allow classes to have multiple inheritance relationships defined by *rdfs : subClassOf*. The Resource Description Framework (RDF) is a general-purpose language for representing information on the Web and RDFS is an RDF schema. A subclass inherits all the properties of its superclass and can extend the superclass with its own definitions. Generically, the subclass has more restrictions than its superclass. Because of the generalization relationship, an instance in the subclass should also belong to the superclass.

- **Collection computing.** Based on set theory, OWL defines the correlations among ontology classes, such as:
  - **Equivalence.** Equivalent classes must have precisely the same instances. That is, for any two classes C_1 and C_2 and an instance c, if $C_1 \equiv C_2$ and $c \in C_1$, it implies that $c \in C_2$, and vice versa. OWL uses the *owl : equivalent* constructor to declare the equivalence relationship.

  - **Disjointness.** The disjointness of a set of classes guarantees that an individual of one class cannot simultaneously be an instance of another specified class. That is, for any two classes C_1 and C_2, if they are two disjoint classes, then $C_1 \cap C_2 = \emptyset$. OWL uses the *owl : disjointWith* constructor to declare the disjointness relationship.

- **Containment.** OWL supports complex class definitions using set operations, such as intersection, union and complementarity. The constructors for the set operation can be combined and nested in complex class definitions.

In addition, the property of an ontology class can also be defined as classes and have the relationships listed above.

A dependency graph is defined to model the dependency relationships among the ontologies. In this directed graph, a node represents a class of ontology or property and a link represents a dependency between two classes. Different types of links are defined to represent various types of dependency relationships. Figure 6.2 illustrates an example dependency graph.

6.4.1.2 Ontology usage analysis

Ontologies can be instantiated to define the inputs, outputs, operation, process and collaboration of services (see Chapters 3, 4, 5 and Tsai *et al.*, 2007). An

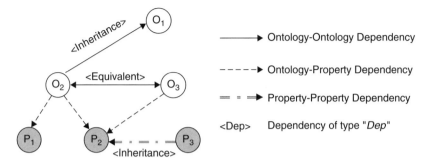

Figure 6.2   Ontology dependency graph.

ontology class can be misused in many ways, such as different scope, restrictions, properties and relationships, which may cause a failure of the service-based software. It is necessary to trace how an ontology is used in a diversified context so as to facilitate software analysis, quality control and maintenance. For example, in case an error is detected in an ontology definition, all the affected atomic and composite services can be located by tracing the usage of the ontology in those software artefacts.

Given an ontology domain D, we define $Ont(a) = \{o_i\}$ as the set of ontology classes $\{\emptyset_i | o_i \in D\}$ used in an artefact a. $Art(o) = \{a_i\}$ is the set of service artefacts that are affected by an ontology class o, $o \in D$, and a_i could be any type of service artefacts such as message, operation, interface, service endpoint and workflow.

Figure 6.3 gives an example to show the usage of an ontology class in different service contexts. In this example, ontology classes are defined for the publication domain. A class *Book* is defined as a subclass of *Publication*. Different *Book-Store* services may use the *Book* ontology to specify the input parameters of the operation *BookDetails()* in the interface *BookQuery*. An application builder defines a business process *BookPurchase* as a workflow of services. In the workflow, it first checks the prices of the book from different *BookStore* services, then orders the book from one with a lower price, and finally checks out the order from the selected service. We can see from the example that a domain model can be used by various services and workflows. Therefore, the quality of the domain model has significant impacts in the services and applications in the domain.

6.4.1.3   Service workflow analysis

A composite service defines the workflow of a set of constituent services. For example, in OWL-S, the ServiceModel is modelled as a workflow of processes including atomic, simple and composite processes. Each composite process holds a *ControlConstruct* which is one of the following: *Sequence, Split, Split-Join, Any-Order, Iterate, If-Then-Else* and *Choice*. The constructs can contain each other recursively, so the composite process can model all of the possible workflows of WS.

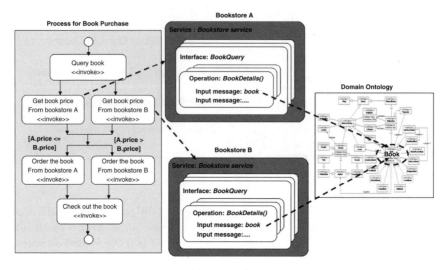

Figure 6.3   Ontology usage.

The risk of the composite service depends on that of each constituent services and the control constructs over them. A service may be conditioned (e.g. *If-Then-Else*) or unconditioned (e.g. *Sequence*) and executed based on the control constructs. The contribution of each constituent is proportional to its execution probability in the composition. In an unconditioned construct, each constituent service will be executed with probability 1; while in a conditioned one, each conditional branch has a certain probability to execute. The higher the path executed, the more the services in the path contribute to the whole composition (Wang *et al.*, 2009).

6.4.2   Failure probability estimation

The failure probability of the service-based system is estimated as follows:

1. Estimate the initial failure probability of each ontology class in the domain.

2. Adjust the estimation of each class by taking its dependencies into consideration.

3. Estimate the initial failure probability of an interface service.

4. Adjust the service's estimation by taking into consideration the failure probability of its input/output parameters defined by the domain ontology.

5. Given the failure probability of a set of functions and their input/output parameters, estimate the failure probability of the composite service based on its control construct analysis.

Here, we use Bayesian statistics to analyse the initial failure probability p of each ontology class and service function. Given a certain artefact a, suppose that we test a for n times with x times failure. Suppose that X is the distribution of x and assume that X follows a binomial distribution, that is $X : b(n, p)$. Assume that the failure probability follows a uniform distribution, that is $p : U(0, 1)$, and its density function is $\pi(p) = 1(0 < p < 1)$. Then, the posterior probability $P(p|X)$ follows $\beta(x + 1, n - x + 1)$ and the expectation of the posterior with respect to p is the estimator of the initial failure probability as follows:

$$E_{P(p|X)}(p) = \frac{x + 1}{n + 2} \qquad (6.4)$$

6.4.2.1 Ontology analysis

The initial failure probability is adjusted with the ontology dependency relationships. The dependence graph (DG) is transformed into a Bayesian network (BN) for inference of the probability of each class. The nodes in the BN represent the ontology classes and the links represent the dependency relationships. Depending on the types of class dependencies, the transformation rules are defined as follows:

1. An ontology class in DG is mapped directly to a node in a BN.

2. The property classes in DG are not shown in the BN. However, the failure probabilities of those property classes are used to estimate the ontology classes that the property belongs to. As an error in a property that will cause an error in the ontology class, we use the product of all the properties as the affecting factor. The adjusted failure probability of the ontology class is definied as follows:

$$P_{adj}(o) = 1 - (1 - P(o)) \times \prod_{j=1..n} (1 - P(p_j)) \qquad (6.5)$$

   where $P(o)$ is the estimated failure probability of the ontology o and P_{adj} is the adjusted probability taking the relationships into considerations; and $p_j \in Prop(o)$, where $Prop(o)$ is the set of properties of o of length n and $P(p_j)$ is the failure probability of the property class p_j.

3. For two classes with the *Inheritance* relationship, that is $Inherit(o_1, o_2)$ where o_1 is the parent of o_2, a directed link is added to BN between o_1 and o_2 starting from o_1 and ending at o_2 to denote that o_2 is affected by o_1.

4. For two classes with the *CollectionComputing* relationship, that is $Equiv(o_1, o_2)$ or $Disjoin(o_1, o_2)$, then (a) two nodes $o_{1'}$ and $o_{2'}$ are added to BN with $P(o_1) = P(o_{1'})$ and $P(o_2) = P(o_{2'})$; and (b) two links are added from o_1 to $o_{2'}$ and from o_2 to $o_{1'}$ to denote the mutual dependence relationship.

Once the BN is created, the standard BN formulae can be used to calculate the probabilities as follows:

$$P(o_i|E_c) = \frac{P(o_i, E_c)}{P(E_c)} \tag{6.6}$$

where E_c is the current evidence (or the current observed nodes) and o_i is the node of ontology class.

6.4.2.2 Service analysis

The failure probability of a service is calculated by multiplying that of its functions and its ontologies, as follows:

$$P(s) = 1 - (1 - P_f(s)) \times \prod_{i=1..n} (1 - P_{adj}(o_i)) \tag{6.7}$$

where:

- $P(s)$ is the failure probability of a service.

- $P_f(s)$ is the failure probability of service functionality.

- $o_i \in Ont(s)$ is the set of ontology classes used in the service definition.

- $P_{adj}(o_i)$ is the adjusted failure probability of each ontology class.

6.4.2.3 Composite service analysis

The failure probability of the composite service is based on its control construct. For unconditioned execution, the failure of each service in the construct will result in failure of the construct; hence the product formula is used to calculate the construct failure probability. For a conditioned construct, we use the weighted sum formula where the weight denotes the execution probability of each branch where the service is located.

We use cc to denote the control construct, and $S(cc) = \{s_i\}$ is the set of services in cc. ρ_i is the execution probability of a service s_i and $\sum \rho_i = 1$. Table 6.1 shows the formulae to calculate.

6.4.3 Importance estimation

We use importance measurements to estimate service failure consequences. A failure in important services may result in a high loss, thus the importance of a service implies the severity level of its failures. The importance of an element is evaluated from two perspectives:

1. Based on dependence analysis, that is the more an element is dependent upon by others, the more important it is.

2. Based on usage analysis, that is the more an element is used in various contexts, the more important it is.

Table 6.1   Failure probability of control construct.

Category	Graphic expression	OWL-S example	$P(cc)$
Unconditioned		Sequence, Split, Any-Order	$P(cc) = 1 - \prod_{i=1,n} (1 - P(S_i))$
Conditioned		If-Then-Else, Choice, Iterate, While-Repeat	$P(cc) = \sum \rho_i P(S_i)$

6.4.3.1 Dependence-based estimation

Given a domain D, the importance of ontology class o is calculated as a weighted sum of the number of its dependent classes, including both directed and indirected dependent classes. For any two classes o_1 and o_2, if there is a path between them in DG, we define the distance $Dis(o_1, o_2)$ between them as the length of links between them. Assume that there exists at least one dependence relationship in the domain, that is $\exists o_1, o_2 \in D$ such that $Dep(o_1, o_2)$. Then, the dependence-based importance estimation of an ontology class is calculated as follows:

$$D_{da}(o) = \sum e^{1-i} |Dep_i(o)| \tag{6.8}$$

$$D_{dr}(o) = \frac{C_{da}(o)}{max_D C_{da}(o_j)} \tag{6.9}$$

where:

- $o \in D$ is an ontology class in the domain and $C_{da}(o)$ is the absolute importance of o, while $C_{dr}(o)$ is the relative importance of o using the dependence-based approach.

- $Dep_i(o) = o_j$ is the set of ontology classes that are dependent upon o with distance i, that is $o_j \in D$, $Dep(o, o_j)$ and $Dis(o, o_j) = i$.

- $|Dep_i(o)|$ is the length of the set, that is the number of dependent ontology classes.

- $max_D C_a(o_j)$ is the maximum absolute importance $\forall o_j \in D$.

6.4.3.2 Usage-based estimation

As shown in Figure 6.3, an ontology class can be instantiated in various services and a service can be integrated in various business processes. The usage model tracks how an ontology or a service is used in different contexts and measures the importance of the element as a weighted sum of the count of the context. Suppose that, for an element e (e could be an ontology class or a service), $Context(e) = \{ct_i\}$ is the set of contexts that e is used in. Assume that the element is used in at least one context for at least once. Then the importance can be measured as follows:

$$C_u(e) \frac{\sum_{i=1}^{|Context(e)|} w_i Num(e, ct_i)}{\sum_{i=1}^{|Context(e)|} Num(e, ct_i)} \qquad (6.10)$$

where

- $C_u(e)$ is the importance of e using the usage-based approach.
- $|Context(e)|$ is the number of contexts that e is used in.
- w_i is the weight for a context $ct_i \in Context(e)$.
- $Num(e, ct_i)$ is the number of e's usage in a context ct_i.

Given a large number of measured elements, we can also use the static models to normalize the values, such as the Bayesian model for collective choice, as follows:

$$C_{ub}(e) = \frac{\frac{1}{N} \sum_{i=1}^{N} C_u(e_i) + C_u(e)}{\frac{1}{N} \sum_{i=1}^{N} |contxt(e_i)| + |Context(e)|} \qquad (6.11)$$

where

- $E = \{e_i\}$ is the set of measured elements.
- $N = |E|$ is the number of measured elements.
- $|Context(e)|$ is the number of contexts that an element e is used in.

6.5   Risk-based adaptive group testing

A key technique of WS group testing is to rank the test cases and exercise them progressively. However, as testing progresses, changes can occur in the service-based system in various ways:

- The services could change as the providers maintain the software, update its functionalities and improve its quality.

- The service composition could change. The application builder may select different service providers for a constituent service and may change the workflow of the composite service to meet the changed requirements of business processes.

- The risks of the software could change due to changes in services and service compositions.

- The potency of the test cases could change due to changes in services and service compositions.

- The quality preference could change. For example, for a safety-critical or mission-critical usage context, it may be required to have a comprehensive coverage of test cases for a service to be accepted; while, otherwise, less strict criteria can be used to reduce the time and cost of testing.

To accommodate these changes, adaptation is necessary to adjust continuously the measurement of software and test cases, and the rules for test cases selection, prioritization and service evaluation. In the proposed risk-based approach, a dynamic mechanism is introduced in order to enable adaptive risk assessment and test case ranking based on the runtime monitoring and profiling of the target services and systems.

6.5.1 Adaptive measurement

With the support of runtime monitoring, information can be gathered on the ontology dependencies, services usage and service workflows. Profiling and statistical analysis of the logged information can facilitate the detection of changes in the system and adjust the measurement of risks and test case potencies.

For example, in an experiment, a sensor is inserted in the process engine of a composite service (Bai *et al.*, 2008) to monitor service calls. The number and sequence of calls to each external service are recorded. Table 6.2 lists the typical sequence of service invocations and the number of invocations to each service in the three time intervals. Observation of the logged data show that:

- In the interval $[t_1, t_2]$, s_2 and s_3 are conditionaly executed after s_1 and the execution probability of each service is $\rho_{s_2} = 0.8$ and $\rho_{s_3} = 0.2$. s_4 and s_5 are executed in parallel after s_2. s_6 is executed after s_3.

- In the interval $[t_2, t_3]$, service s_5 becomes unavailable and there is no subsequent invocation of s_5 after s_2. In addition, the call distribution between s_2 and s_3 is changed from 0.8:0.2 to 0.5:0.5.

- In the interval $[t_3, t_4]$, service s_7 is newly bound to the system and invoked after s_2 in parallel to s_4. The call distribution between s_2 and s_3 is resumed to 0.7:0.3.

Table 6.2   Example service composition profile (number of invocations in the profile).

Time interval	Execution sequence	s_1	s_2	s_3	s_4	s_5	s_6	s_7
$[t_1, t_2]$	$\{s_1, s_2, s_4, s_5\}$	100	80	20	80	80	20	–
	$\{s_1, s_2, s_5, s_4\}$							
	$\{s_1, s_3, s_6\}$							
$[t_2, t_3]$	$\{s_1, s_2, s_4\}$	50	25	25	25	–	25	–
	$\{s_1, s_3, s_6\}$							
$[t_3, t_4]$	$\{s_1, s_2, s_4, s_7\}$	200	140	60	140	–	60	140
	$\{s_1, s_2, s_7, s_4\}$							
	$\{s_1, s_3, s_6\}$							

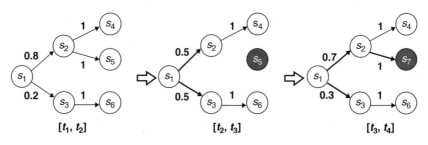

Figure 6.4   Example service composition changes.

The changing process of the composition structure is shown in Figure 6.4. Similar to the monitored composition changes, the system can also detect the changes in the ontology domain and ontology usage. Such changes trigger a reassessment of the two factors of risk: failure probability and importance. Considering this example, Table 6.3 shows the changes in the risk of the composite service.

Table 6.3   Example of adaptive risk assessment.

Time interval	Risk factors	s_1	s_2	s_3	s_4	s_5	s_6	s_7	Composite service
$[t_1, t_2]$	P(s)	0.1	0.3	0.1	0.3	0.7	0.3	–	0.654
	C(s)	1	0.8	0.2	0.3	0.3	0.3	–	
$[t_2, t_3]$	P(s)	0.1	0.3	0.1	0.3	–	0.3	–	0.840
	C(s)	1	0.5	0.5	0.5	–	0.5	–	
$[t_3, t_4]$	P(s)	0.1	0.3	0.1	0.3	–	0.3	0.3	0.751
	C(s)	1	0.7	0.3	0.3	–	0.3	0.3	

6.5.2 Adaptation rules

Rules are defined to control the testing process. In the generic WS group testing process, rules are defined to control the following testing activities:

- Risk levels to categorize the test cases and arrange them hierarchically into different layers.

- The strategies of ranking the test cases, such as cost, potency, criticality, dependency, etc.

- The strategies for ruling out services after each layer of testing.

- The strategies of ranking the services, such as importance, failure rates on the test case, etc.

- The entry and exit criteria for each layer of group testing.

In this application, testing is controlled as a risk mitigation process. That is, the test cases that have a high probability to detect a risky bug should be exercised first so that the risky bugs can be detected and removed early and the risk of the whole system can be reduced. As bugs are detected and removed, the rules of the strategies and criteria are also adapted to reflect the changes in the risks of the services. The rule adaptation is by nature a problem of dynamic planning. Each layer in the WS progressive group testing is a stage in decision making, and the goal is to select a set of test case with maximum potential risks.

6.6 Evaluation

Suppose that $T = \{t_i\}$ is the set of test cases, $TE = \{t_i\} \subseteq T$ is the set of exercised test cases, $B = \{bug_i\}$ is the set of all bugs in the system, and $BD = bug_i \subseteq B$ is the set of bugs that has been detected. To evaluate the test results, we define the following two metrics:

1. **Test cost.** That is, the average number of test cases required to detect a bug in the system:

$$Cost(T) = \frac{|TE|}{|BD|} \qquad (6.12)$$

2. **Test efficiency.** That is, given a number of exercised test cases, the ratio between the percentage of bugs detected and the percentage of test cases exercised:

$$Effect(T, TE) = \frac{|BD|}{|B|} \Big/ \frac{|TE|}{|T|} \qquad (6.13)$$

The goal of testing is to detect as many bugs as possible with as few as possible test cases (Myers, 1979). That is, low test cost and high test efficiency are preferred.

Two evaluation experiments were exercised on case studies. For simplicity, we only consider the bugs in the ontology classes. Assume that each ontology class has exactly one bug, which could be detected by one test case. The test cases are categorized based on their target ontology classes. Each ontology class is assigned 200 test cases. All the test cases are independent and they are designed for exactly one ontology class.

In Experiment 1, eight ontology classes are identified; the corresponding BN is shown in Figure 6.5. We simulate the joint distribution of the BN. Table 6.4 shows the calculated risk of each node during the iterations of nodes risk analysis and test selection. Figure 6.6 shows the results of the experiment which compares the proposed risk-based approach (RBT) with the random testing approach (RT). In the figure, the horizontal axis lists the number of bugs detected. The curves show the trend of test cost (Figure 6.6(a)) and test efficiency (Figure 6.6(b)) with increasing number of bugs detected. Here, the RT approach is estimated with two probability levels, 80% and 90%; that is, to detect bugs with a certain probability and the number of test cases required.

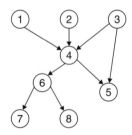

Figure 6.5   The Bayesian network of experiment 1.

In Experiment 2, 16 ontology classes are identified and the corresponding BN is shown in Figure 6.7. Taking experiment 1 as a training process for experiment 2, we introduce the adaptation mechanism to rank test cases based on their defect detection history. Figure 6.8 shows the experimental results which compare test cost (Figure 6.8(a)) and test efficiency (Figure 6.8(b)) with increasing number of bugs detected.

From the experiment we can see that the risk-based approach can greatly reduce the test cost and improve test efficiency. The learning process and the adaptation mechanism can greatly improve the test quality.

6.7   Summary

Testing is critical to assure the quality properties of a service-based system so that services can deliver their promise. To reduce test cost and improve test

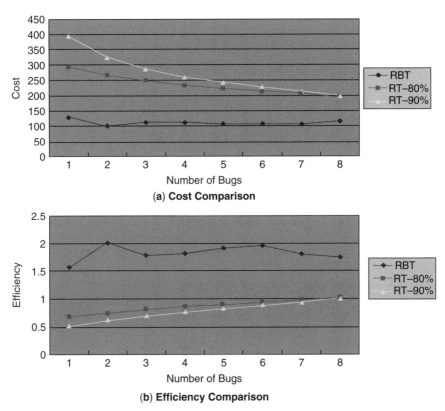

Figure 6.6   Comparison of test costs and test effectiveness between RBT and RT of experiment 1.

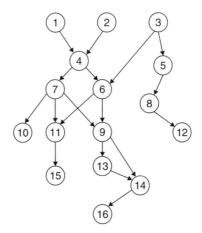

Figure 6.7   The Bayesian network of experiment 2.

Table 6.4   Adaptive risk assessment of the ontology classes in experiment 1.

Nodes	1	2	3	4	5	6	7	8
Round 1	0.3000	0.5000	0.8000	0.5300	0.4130	0.8060	0.7612	0.8418
Round 2	0.3015	0.5036	0.8000	0.5478	0.4148	0.8617	0.7723	Observed
Round 3	0.3052	0.5124	0.8000	0.5918	0.4192	observed	0.8000	Observed
Round 4	0.3052	0.5124	0.8000	0.5918	0.4192	Observed	Observed	Observed
Round 5	0.3052	0.5124	Observed	Observed	0.4592	Observed	Observed	Observed
Round 6	0.3369	0.5943	Observed	Observed	0.4130	Observed	Observed	Observed
Round 7	0.3333	Observed	Observed	Observed	0.4130	Observed	Observed	Observed
Round 8	0.3333	Observed	Observed	Observed	Observed	Observed	Observed	Observed

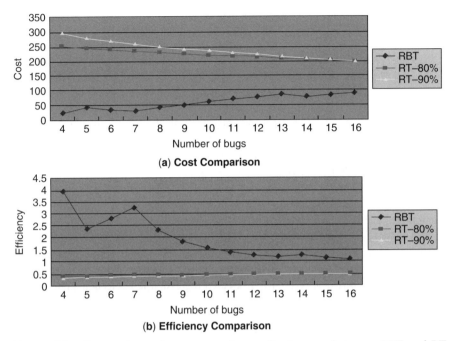

Figure 6.8   Comparison of test cost and test effectiveness between RBT and RT of experiment 2.

efficiency, we propose a risk-based approach to ranking and selecting test cases for controlling the process of WS group testing. An adaptation mechanism is introduced so that the risks can be dynamically measured and the control rules can be dynamically adjusted online. Motivated by the unique testing issues, we show the convergence among various disciplines including statistical, service-oriented computing, and semantic engineering. Some preliminary results illustrate the feasibility and advantages of the proposed approach. Relevant future areas of work include: (1) application and experiments with a real application domain, such as envisaged by the MUSING project (MUSING, 2006); (2) service behaviour analysis to identify the typical fault models and risky scenarios; (3) model enhancement to achieve better test evaluation results by using more sophisticated mathematical models and reasoning techniques. For more examples of system and process improvement see Kenett and Baker (1999, 2010).

References

Amland, S. (2000) Risk-based testing: risk analysis fundamentals and metrics for software testing including a financial application case study, *Journal of Systems and Software*, 53, 3, pp. 287–295.

Bach, J. (1999) Heuristic risk-based testing, *STQE Magazine*, 1, 6.

Bai, X. and Kenett, R.S. (2009) Risk-based adaptive group testing of Web Services, *COMPSAC (IWSC)*, 2, pp. 485–490.

Bai, X., Cao, Z. and Chen, Y. (2007a) Design of a trustworthy service broker and dependence-based progressive group testing, *International Journal of Simulation and Process Modeling*, 3, 1/2, pp. 66–79.

Bai, X., Chen, Y. and Shao, Z. (2007b) Adaptive Web Services testing, *IWSC*, pp. 233–236.

Bai, X., Lee, S., Liu, R., Tsai, W.T. and Chen, Y. (2008) Collaborative Web Services monitoring with active service broker, *COMPSAC*, pp. 84–91.

Berners-Lee, T., Handler, J. and Lassila, O. (2008) The Semantic Web, *Scientific American Magazine*, May 2001, revised.

Cai, K.Y. (2002) Optimal software testing and adaptive software testing in the context of software cybernetics, *Information and Software Technology*, 44, pp. 841–844.

Canfora, G., and Penta, M. (2009) *Service-Oriented Architectures Testing: A Survey*, Lecture Notes in Computer Science, Vol. 5413, Springer-Verlag, pp. 78–105.

Cangussu, J.W., Miller, S.D., Cai, K.Y. and Mathur, A.P. (2009) Software cybernetics, in *Encyclopaedia of Computer Science and Engineering*, John Wiley & Sons, Ltd.

Chen, Y. and Probert, R.L. (2003) A risk-based regression test selection strategy, *14th ISSRE*.

Dorfman, R. (1964) The detection of defective members of large population, *Annals of Mathematical Statistics*, 14, pp. 436–440.

Finucan, F.M. (1964) The blood testing problem, *Applied Statistics*, 13, pp. 43–50.

Gomez-Perez, A., Fernandez-Lopez, M. and Corcho, O. (2005) *Ontological Engineering*, Springer-Verlag.

Google Map API (2010) http://code.google.com/intl/zh-CN/apis/maps/ (11 January 2010).

Harel, A., Kenett, R.S. and Ruggeri, F. (2008) Decision support for user interface design: usability diagnosis by time analysis of the user activity, *COMPSAC*.

Karolak, D.V. (1996) *Software Engineering Risk Management*, IEEE Computer Society Press.

Kenett, R.S. and Baker, E. (1999) *Software Process Quality: Management and Control*, Marcel Dekker (Taylor & Francis, Kindle Edition, 2007).

Kenett, R.S. and Baker, E. (2010) *Process Improvement and CMMI for Systems and Software: Planning, Implementation, and Management*, Auerbach Publications.

Kenett, R.S. and Pollak, M. (1986) A semi-parametric approach to testing for reliability growth with an application to software systems, *IEEE Transactions on Reliability*, R-35, 3, pp. 304–311.

Kenett, R.S. and Steinberg, D. (2006) New frontiers in design of experiments, *Quality Progress*, August, pp. 61–65.

Kenett, R.S. and Tapiero, C. (2010) Quality and risk: convergence and perspectives, *Risk and Decision Analysis*, 4, 1, pp. 231–246.

Kenett, R.S. and Zacks S. (1998) *Modern Industrial Statistics: Design and Control of Quality and Reliability*, Duxbury Press.

Kenett, R.S., Harel, A. and Ruggeri, F. (2009) Controlling the usability of web services, *International Journal of Software Engineering and Knowledge Engineering*, 19, 5, pp. 627–651.

Martin, D. (2004) Bringing semantics to Web Services: the OWL-S approach, *SWSWPC*.

MUSING (2006) IST- FP6 27097, http://www.musing.eu (accessed 21 May 2010).

Myers, G. (1979) *The Art of Software Testing*, John Wiley & Sons, Ltd.

O'Reilly, T. (2005) What is Web 2.0?, http://oreilly.com/web2/archive/what-is-web-20.html (accessed 21 May 2010).

Ottevanger, I. (1999) A risk-based test strategy, *STARWest*.

Papazoglou, M.P. and Georgakopoulos, D. (2003) Service-oriented computing, *Communications of the ACM*, 46, 10, pp. 25–28.

Redmill, F. (2004) Exploring risk-based testing and its implications, *Software Testing, Verification and Reliability*, 14, pp. 3–15.

Rent API (2010) http://www.programmableweb.com/api/rentrent (accessed 6 March 2010).

Rosenberg, L.H., Stapko, R. and Gallo, A. (1999) Risk-based object oriented testing, *24th SWE*.

Singh, M.P. and Huhns, M.N. (2005) *Service-Oriented Computing*, John Wiley & Sons, Ltd.

Sobel, M. and Groll, P.A. (1959) Group testing to eliminate all defectives in a binomial sample, *Bell System Technical Journal*, 38, 5, pp. 1179–1252.

Spies, M. (2009) An ontology modeling perspective on business reporting, *Information Systems*, 35, 4, pp. 404–416.

Taleb, N.N. (2007) *The Black Swan: The impact of the highly improbable*, Random House.

Taleb, N.N. (2008) The fourth quadrant: a map of the limits of statistics, *Edge*, http://www.edge.org/3rd_culture/taleb08/taleb08_index.html (accessed 21 May 2010).

Tsai, W.T., Paul, R., Cao, Z., Yu, L., Saimi, A. and Xiao, B. (2003a) Verification of Web Services using an enhanced UDDI server, *Proceedings of IEEE WORDS*, pp. 131–138.

Tsai, W.T., Paul, R., Yu, L., Saimi, A. and Cao, Z. (2003b) Scenario-based web services testing with distributed agents, *IEICE Transactions on Information and Systems*, E86-D, 10, pp. 2130–2144.

Tsai, W.T., Chen, Y., Cao, Z., Bai, X., Huang, H. and Paul, R. (2004) Testing Web Services using progressive group testing, *Advanced Workshop on Content Computing*, pp. 314–322.

Tsai, W.T., Bai, X., Chen, Y. and Zhou, X. (2005) Web Services group testing with windowing mechanisms, *SOSE*, pp. 213–218.

Tsai, W.T., Huang, Q., Xu, J., Chen, Y. and Paul, R. (2007) Ontology-based dynamic process collaboration in service-oriented architecture, *SOCA*, pp. 39–46 2007.

W3C (2010a) OWL Web Ontology Language Overview, http://www.w3.org/TR/2004/REC-owl-features-20040210/ (11 January 2010).

W3C (2010b) OWL-S: Semantic Markup for Web Services, http://www.w3.org/Submission/2004/SUBM-OWL-S-20041122/ (11 January 2010).

Wang, L., Bai, X., Chen, Y. and Zhou, L. (2009) A hierarchical reliability model of service-based software system, *COMPSAC*, 1, pp. 199–208.

Watson, G.S. (1961) A study of the group screening method, *Technometrics*, 3, 3, pp. 371–388.

Part III

OPERATIONAL RISK ANALYTICS

7

Scoring models
for operational risks

Paolo Giudici

7.1 Background

The motivation of this chapter is to present efficient statistical methods aimed at measuring the performance of business controls, through the development of appropriate operational risk indicators. Recent legislation and a number of market practices are motivating such developments, for instance the New Basel Capital Accord known as Basel II (Basel Committee on Banking Supervision, 2001) published by the Basel Committee on Banking Supervision and internationally adopted by supervisory authorities.

In the context of information systems, the recently developed ISO 17799 (ISO, 2005) establishes the need for risk controls aimed at preserving the security of information systems. Finally, the publicly available specification PAS56 (PAS, 2006), in setting criteria that should be met to maintain business continuity of IT-intensive companies, calls for the development of statistical indicators aimed at monitoring the quality of business controls in place.

The focus of the chapter is on the Basel II Accord, keeping in mind that what is developed here for the banking sector can be extended to the general enterprise risk management framework (see Chapter 1 and Bonafede and Giudici, 2007; Figini and Giudici, 2010).

Operational Risk Management: A Practical Approach to Intelligent Data Analysis   Edited by Ron S. Kenett
and Yossi Raanan   © 2011 John Wiley & Sons, Ltd

The Bank of International Settlements (BIS) is an international financial institution whose main purpose is to encourage and facilitate cooperation among central banks. In particular, BIS established a commission, the Basel Committee on Banking Supervision (BCBS), to formulate broad supervisory standards and guidelines and to recommend statements of best practice. The ultimate purpose of the committee is the prescription of capital adequacy standards for all internationally active banks. In 1988 the BCBS issued one of the most significant international regulations with an impact on the financial decision of banks: the Basel Accord. Subsequently, the BCBS worked on a revision, called the New Accord on Capital Adequacy, or Basel II (Basel Committee on Banking Supervision, 2001). This new framework was developed to ensure the stability and soundness of financial systems. It is based on three 'pillars': minimum capital requirements, supervisory review and market discipline.

The novelty of the new agreement was the identification of operational risk as a new category separated from the others. In fact, it was only with the new agreement that the Risk Management Group of the Basel Committee proposed the current definition of operational risk: 'Operational Risk is the risk of loss resulting from inadequate or failed internal processes, people and systems or from external events.'

The Risk Management Group also provided a standardized classification of operational losses into eight business lines (BLs) and seven event types (ETs). For more information on Basel II see Chapter 10.

The aim of operational risk measurement is twofold: on the one hand, there is a risk contingency aspect which involves setting aside an amount of capital requirements that can cover unexpected losses. This is typically achieved by estimating a loss distribution and deriving from it specific functions of interest (such as value at risk). On the other hand, there is the managerial need to rank operational risks in an appropriate way, say from high priority to low priority, so as to identify appropriate management actions directed at improving preventive controls on such risks (Alexander, 2003; King, 2001; Cruz, 2002).

In general, the measurement of operational risks leads to the measurement of the efficacy of the controls in place at a specific organization: the higher the operational risks, the worse such controls.

The complexity of operational risks and the newness of the problem have driven international institutions, such as the Basel Committee, to define conditions that sound statistical methodologies should satisfy in order to build and measure adequate operational risk indicators.

7.2   Actuarial methods

Statistical models for operational risk management (OpR) are grouped into two main categories: 'top-down' and 'bottom-up' methods. In the former, risk estimation is based on macro data without identifying the individual events or the causes of losses. Therefore, operational risks are measured and covered at a central level, so that local business units are not involved in the measurement and allocation process.

'Top-down' methods include the basic indicator approach (see e.g. Yasuda, 2003; Pezier, 2002) and the standardized approach, where risk is computed as a certain percentage of the variation of some variable, such as gross income, considered as a proxy for the firm's performance (Cornalba and Giudici, 2004; Pezier, 2002). This first approach is suitable for small banks, which prefer an inexpensive methodology that is easy to implement.

'Bottom-up' techniques use individual events to determine the source and amount of operational risk. Operational losses can be divided into separate levels that correspond to business lines and event types and risks are measured at each level and then aggregated. These techniques are particularly appropriate for large banks and banks operating at the international level. It requires implementation of sophisticated methods that are tailored to the bank's risk profile. Methods belonging to this class are grouped into the advanced measurement approach (AMA) (BCBS, 2001). Under AMA, the regulatory capital requirement is equal to the risk measure generated by the bank's internal operational risk measurement system using the quantitative and qualitative criteria set by the Basel Committee. It is an advanced approach as it allows banks to use external and internal loss data as well as internal expertise (Giudici and Bilotta, 2004).

Statistical methods for operational risk management in the bottom-up context have been developed only recently. One main approach that has emerged in this areas is the actuarial approach. The method is applicable in the presence of actual loss data, and is based on the analysis of all available and relevant loss data with the aim of estimating the probability distribution of the losses. The most common methods described (e.g. King, 2001; Cruz, 2002; Frachot et al., 2001; Dalla Valle et al., 2008) are often based on extreme value distributions. Another line of research suggests the use of Bayesian models for estimating loss distributions (see Yasuda, 2003; Cornalba and Giudici, 2004; Fanoni et al., 2005; Figini and Giudici, 2010).

The main disadvantage of actuarial methods is that their estimates rely only on past data, thus reflecting a backward-looking perspective. Furthermore, it is often the case, especially for smaller organizations, that, for some business units, there are no loss data at all. Regulators thus recommend developing models that can take into account different data streams, not just internal loss data (BCBS, 2001). These streams may be self-assessment opinions, usually forward looking; external loss databases, usually gathered through consortiums of companies; and data on key performance indicators.

In the actuarial model, loss events are assumed to be independent and, for each of them, it is assumed that the total loss in a given period (e.g. one year) is obtained as the sum of a random number (N) of impacts (X_i). In other words, for the jth event the loss is equal to

$$L_j = \sum_{i=1}^{N_j} X_{ij}$$

Usually the distribution of each j-specific loss is obtained from the specification of the distribution of the frequency N and the mean loss or *severity* S. The convolution of the two distributions leads to the distribution of loss L (typically through a Monte Carlo estimation step), from which a functional of interest, such as the 99.9% percentile, the value at risk (VaR), can be derived (see Chapter 2). In other words, for each risk event j, we consider the sum of the losses of the N_j events that occur over the time horizon considered, possibly by units one year long.

7.3   Scorecard models

The scorecard approach is based on *self-assessment*, which is based on the experience and opinions of a number of internal 'experts' of the company, who usually correspond to a particular business unit. An internal procedure of control self-assessment can be periodically done through questionnaires, submitted to risk managers (experts), which provide information such as the quality of internal and external control systems of the organization on the basis of its own experience in a given period. In a more sophisticated version, experts can also assess the frequency and mean severity of the losses for such operational risks (usually in a qualitative way).

Self-assessment opinions can be summarized and modelled so to attain a ranking of the different risks and a priority list of interventions in terms of improvement of the related controls.

In order to derive a summary measure of operational risk, perceived losses contained in the self-assessment questionnaire can be represented graphically (e.g. through a histogram representation) and lead to an empirical non-parametric distribution. Such a distribution can be employed to derive a functional of interest, such as the 99.9% percentile (VaR).

Scorecard models are useful for prioritizing interventions on the control system, so as to reduce effectively the impact of risks, *ex ante* and not a posteriori, as can be done by allocating capital (corresponding to the VaR).

A methodology aimed at summarizing concisely and effectively the results of a self-assessment questionnaire is presented below. We present the methodology in the context of a real-case application.

Suppose that we are given 80 risk events (this is the order of magnitude employed in typical banking operational risk management analysis). These events can be traced to the four main causes of operational risk: people, processes, systems and external events (see Chapter 3).

The assessment is based on the opinions of a selected number of banking professionals (both from headquarters and local branches). The aim of the assessment questionnaire is first described in a group presentation. Following the presentation, a pilot draft questionnaire is conducted. The nature and structure of each risk-related question is clarified in a focus group discussion with managers of the bank.

The result of this preliminary analysis is that each of the selected professional is asked, for a total of the 80 risk events, his/her opinion on the frequency, severity and effectiveness of the controls in place for each event. The number of possible frequency classes is equal to four: daily, weekly, monthly and yearly. The number of severity classes depends on the size of capital of the bank, with an average of six or seven classes, going from 'an irrelevant loss' to 'a catastrophic loss'. Finally the number of possible classes of controls is three: not effective, to be adjusted, effective.

Once interviews are completed, the aim is to assign a 'rating' to each risk event, based on the distribution of the opinions on the frequency, controls and severity. The approach we propose is to use the median class as a location measure of each distribution, and the normalized Gini index as an indicator of the 'consensus' on such a location measure where the Gini index is equal to

$$G = 1 - \sum_{i=1}^{K} p_i^2$$

where K is the number of classes and p_i the relative frequency of each of such classes. The index can be normalized by dividing it by its maximum value, equal to $1 - 1/K$.

This results in three rating measures for each event, expressed using the conventional risk letters: A for low risk, B for medium risk, C for higher risk, and so on.

While the median is used to assign a 'single-letter' measure, the Gini index is used to double or triple the letter, depending on the value of the index. For example, if the median of the frequency distribution of a certain risk type (e.g. theft and robbery) is 'yearly', corresponding to the lowest risk category, letter A is assigned. Then, if all those interviewed agree on that evaluation (e.g. the Gini index is equal to zero), A is converted to AAA; if instead the Gini index corresponds to maximum heterogeneity A, it remains A. Intermediate cases will receive a double rating of AA.

The same approach can be followed for the severity as well as for the controls, leading to a complete scorecard that can be used for intervention purposes.

For visualization purposes, colours are associated to the letters, using a 'traffic-light' convention: green corresponds to A; yellow to B; red to C; purple to D; and so on.

Figure 7.1 presents the results from the scorecard model, for a collection of risk events belonging to people (internal frauds) and external events (external frauds and losses at material activities).

From Figure 7.1, it turns out that the event 1.2.6 should be given a priority 1 of intervention, as controls are not effective, and both frequency and severity are yellow. Other events at risk include 2.2.1 and 2.2.4 which have a high frequency and medium quality controls. Note that the opinion on the severity is usually considered second in priority determination as it typically concerns a mean value which cannot be modified by the action of controls.

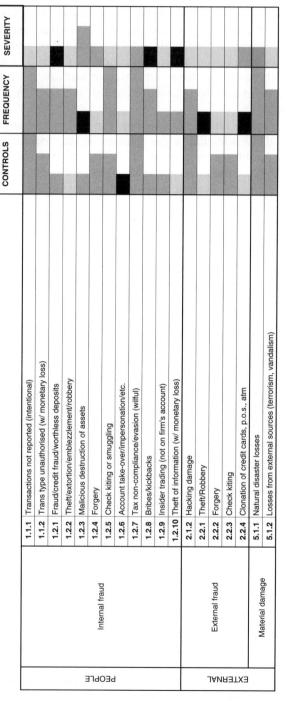

Figure 7.1   Example of results from our proposed scorecard model.

7.4   Integrated scorecard models

While scorecard methods typically use self-assessment data, actuarial models use internal loss data.

The disadvantage of these approaches is that they consider only one part of the statistical information available for estimating operational risks. Actuarial methods rely only on past loss data (backward looking) and, therefore, do not consider important information on the perspective and the evolution of the considered company. On the other hand, scorecard methods are based only on perceived data (forward looking) and, therefore, do not necessarily reflect well past experience.

A further problem is that, especially for rare events, a third data stream may be considered: external loss data. This source of data is made up of pooled records of losses, typically higher than a certain value (e.g. €5000 (about $6000)), collected by an appropriate association of banks.

It therefore becomes necessary to develop a statistical methodology that is able to merge three different data streams in an appropriate way, yet maintaining simplicity of interpretation and predictive power. Here we propose a flexible non-parametric approach that can reach these objectives. Such an approach can be justified within a non-parametric Bayesian context.

Consider, for each event, that all loss data occurred in the past as well as the expected self-assessment losses for the next period. The latter is counted as one data point, typically higher than actual losses, even when it is calculated as a mean loss rather than as a worst case loss.

Putting together the self-assessment data point with the actual loss data points, we obtain an integrated loss distribution, from which VaR can be directly calculated. Alternatively, in order to take the losses of the distributions more correctly into account, a Monte Carlo simulation can be performed on the given losses, leading to a (typically higher) Monte Carlo VaR, parallel to what is usually done in the actuarial approach.

In Figure 7.2 we compare, for a real database, the VaR obtained under a 'pure' self-assessment approach with the actuarial VaRs (both historical and Monte Carlo based) and the integrated (Bayesian) VaR (both simple and Monte Carlo). For reasons of predictive accuracy, we build all methods on a series of data points updated at the end of the year 2005, calculate the VaR for the year 2006 (possibly integrating it with the self-assessment available opinions for 2006) and compare the VaR with the actual losses for 2006. We also calculate the VaR that would be obtained under the simple basic indicator approach (BIA), suggested by Basel II for implementation in small and medium-sized banks. The BIA approach amounts to calculating a flat percentage (15%) of a relevant indicator (such as the gross income), without statistical elaborations.

From Figure 7.2, it turns out that both our proposed models (Bayes VaR and Bayes Monte Carlo) lead to an allocation of capital (represented by the VaR) lower than the BIA approach and higher than the observed losses. Although these results are achieved by the actuarial models as well (historical and actuarial

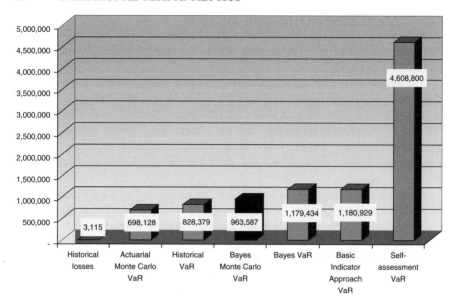

Figure 7.2   Example of results from our integrated scorecard model.

Monte Carlo), we believe that a prudent approach, as presented in this chapter, is a sounder approach, especially over a longer time horizon. For more on such integrated models, see Chapter 8.

7.5   Summary

In this chapter we present different scoring methods for operational risk management that are commonly used. We mention the negative aspects of considering only past event, or considering only future events, ignoring the known history of risk events. We also present a combined method, one that takes into account both the history of risk events and a self-assessment, experts-based, future-looking scorecard. We suggest that this combined method is indeed a better approach for assessing operational risks.

References

Alexander, C. (2003) *Operational Risk: Regulation, Analysis and Management*, Financial Times/Prentice Hall, London.

Basel Committee on Banking Supervision (2001) *Working Paper on the Regulatory Treatment of Operational Risk*, www.bis.org.

Bonafede, E.C. and Giudici, P. (2007) Bayesian networks for enterprise risk assessment, *Physica A*, 382, 1, pp. 22–28.

Cornalba, C. and Giudici, P. (2004) Statistical models for operational risk management, *Physica A*, 338, pp. 166–172.

Cruz, M. (2002) *Modeling, Measuring and Hedging Operational Risk*, John Wiley & Sons, Ltd, Chichester.

Dalla Valle, L., Fantazzini, D. and Giudici, P. (2008) Copulae and operational risk, *International Journal of Risk Assessment and Management*, 9, 3, pp. 238–257.

Fanoni, F., Giudici, P. and Muratori, M. (2005) *Il rischio operativo: misurazione, monitoraggio, mitigazione*, Il sole 24 ore.

Figini, S. and Giudici, P. (2010) Statistical merging of rating models, *Journal of the Operational Research Society*, DOI 10.1057/jors.2010.41.

Frachot, A., Georges, P. and Roncalli, T. (2001) *Loss distribution approach for operational risk*. Working Paper, Groupe de Recherche Opérationnelle du Crédit Lyonnais.

Giudici, P. and Bilotta, A. (2004) Modelling operational losses: a Bayesian approach, *Quality and Reliability Engineering International*, 20, pp. 407–417.

ISO 17799 (2005) www.iso.org/iso/iso_catalogue/catalogue_tc/catalogue_detail.htm? csnumber=39612 (accessed 21 May 2010).

King, J. (2001) *Operational Risk. Measurement and Modelling*, John Wiley & Sons, Ltd, Chichester.

PAS56 (2006) www.pas56.com (accessed 11 January 2010).

Pezier, J. (2002) *A constructive review of the Basel proposals on operational risk*, ISMA Technical Report, University of Reading.

Yasuda, Y. (2003) *Application of Bayesian inference to operational risk management*, Masters Thesis, University of Tsukuba.

8

Bayesian merging and calibration for operational risks

Silvia Figini

8.1 Introduction

The main objective of this chapter is to describe, with a real case study, how to properly integrate financial quantitative data with operational data. The objective is to generate an integrated risk score from two different data sources: financial data represented by XBRL balance sheets and operational loss data. XBRL is a language for the electronic communication of business and financial data, which is revolutionizing business reporting. It provides major benefits in the preparation, analysis and communication of business information and it offers cost savings, greater efficiency and improved accuracy and reliability to all those involved in supplying or using financial data (XBRL, 2010).

The idea behind XBRL is simple. Instead of treating financial information as a block of text – as in a standard Internet page or a printed document – it provides an identifying tag for each individual item of data. This is computer readable. For example, *company net profit* has its own unique tag. The introduction of XBRL tags enables automated processing of business information by computer software, cutting out laborious and costly processes of manual re-entry and comparison. Computers can treat XBRL data 'intelligently', that is they can recognize the information in an XBRL document, select it, analyze it, store it, exchange it with

Operational Risk Management: A Practical Approach to Intelligent Data Analysis Edited by Ron S. Kenett and Yossi Raanan © 2011 John Wiley & Sons, Ltd

other computers and present it automatically in a variety of ways for users. XBRL greatly increases the speed of handling of financial data, reduces the chance of error and permits automatic checking of information.

Companies can use XBRL to save costs and streamline their processes for collecting and reporting financial information. Consumers of financial data, including investors, analysts, financial institutions and regulators, can receive, find, compare and analyze data much more rapidly and efficiently if it is in XBRL format. XBRL can handle data in different languages and accounting standards. It can flexibly be adapted to meet different requirements and uses. Data can be transformed into XBRL by suitable mapping tools or it can be generated in XBRL by appropriate software.

In our case study, starting from XBRL balance sheets, it is possible to derive quantitative information such as financial ratios that are useful for measuring credit risk. Considering the data at hand, as described in Section 8.3, financial ratios are categorized according to the financial aspects of the business which the ratio measures: **liquidity ratios** measure the availability of cash to pay debt; **activity ratios** measure how quickly a firm converts non-cash assets to cash assets; **profitability ratios** measure the firm's use of its assets and the control of its expenses to generate an acceptable rate of return. Finally, **market ratios** measure investor response to owning a company's stock and also to the cost of issuing stock.

On the other hand, operational loss data is typically related to the frequency of a specific event weighted by a corresponding severity (see Chapter 7 and Giudici, 2003; Dalla Valle and Giudici, 2008). In our case study this is a good starting point for measuring operational risk (for more on this issue see Figini *et al.*, 2007; Alexander, 2003; Cruz, 2002). On the basis of the data available, and in order to integrate credit and operational risks, we propose a new methodology. Empirical evidence supporting this approach is provided by a case study with real data. Section 8.2 of this chapter presents our methodological proposal; Section 8.3 describes the application and the main results achieved; and Section 8.4 reports on the conclusions.

8.2   Methodological proposal

In this section we show how to integrate financial and operational risk management and, more precisely, credit risk with operational risk, following a two-step approach. More specifically, we employ logistic regression to derive a quantitative measure of credit risk (probability of default) and unsupervised techniques to derive a quantitative score for operational losses.

For credit risk, we consider a predictive model for a response variable of the qualitative type.

Let y_i, for $i = 1, 2, \ldots, n$, be the observed values of a binary response variable which can take only the values 0 or 1. The level 1 usually represents the occurrence of an event of interest, often called a 'success'.

A logistic regression model is defined in terms of fitted values that are to be interpreted as probabilities that the event occurs, in different subpopulations:

$$\pi_i = P(Y_i = 1), \text{ for } i = 1, 2, \ldots, n$$

More precisely, a logistic regression model specifies that an appropriate function of the fitted probability of the event is a linear function of the observed values of the available explanatory variables, as in the following:

$$\log \left[\frac{\pi_i}{1 - \pi_i} \right] = a + b_1 x_{i1} + b_2 x_{i2} + \cdots + b_k x_{ik}$$

Once π_i is calculated, on the basis of the data, a fitted value for each binary observation, $\hat{y}_i$, can be derived by introducing a cut-off threshold value of π_i above which $\hat{y}_i = 1$ and below which $\hat{y}_i = 0$.

As a final result, we obtain for each company a score π_i, $0 < \pi_i < 1$. The data used in the example below was obtained from SMEs (Small and Medium-sized Enterprises) and therefore the score is assigned to specific SMEs.

Considering operational data, it is often necessary for ease of analysis and description to reduce the dimensionality of the problem as expressed by the number of variables present. The technique that is typically used to achieve this objective is the linear operation known as the principal components transformation (Fuchs and Kenett, 1998; Giudici, 2003).

It is assumed that S, the matrix of the variables, is of full rank; this implies that none of the considered variables is a perfect linear function of the others (or a linear combination of them).

In general, the vth principal component, for $v = 1, \ldots, k$, is the linear combination

$$Y_v = \sum_{j=1}^{p} a_{vj} X_j = X a_v$$

where the vector of the coefficients a_v is the eigenvector of S corresponding to the vth (in order of magnitude) eigenvalue. Such an eigenvector is normalized and orthogonal to all the previously extracted components.

The main difficulty connected with the application of the principal components is their interpretation. This is because they are a linear combination of all the available variables, and therefore they do not have a clear measurement scale. In order to facilitate their interpretation, we will introduce the concepts of absolute importance and relative importance of principal components.

Consider first the absolute importance. In order to solve the maximization problem that leads to the principal components, it can be shown that $Sa_v = \lambda_v a_v$, where λ is the vector of eigenvalues of the matrix S. Therefore, the variance of the vth principal component corresponds to the vth eigenvalue of the data matrix:

$$Var(Y_v) = Var(X a_v) = a_v' S a_v = \lambda_v$$

Concerning the covariance between the principal components, it can be shown that

$$Cov(Y_i, Y_j) = Cov(Xa_i, Xa_j) = a_i' Sa_j = a_i' \lambda_j a_j = 0, \quad i \neq j$$

Because a_i and a_j are assumed to be orthogonal when $i \neq j$, this implies that the principal components are uncorrelated.

The variance–covariance matrix between them is thus expressed by the following diagonal matrix:

$$Var(Y) = \begin{bmatrix} \lambda_1 & & 0 \\ & \ddots & \\ 0 & & \lambda_k \end{bmatrix}$$

Consequently, the following ratio expresses the proportion of variability that is 'maintained' in the transformation from the original p variables to $k < p$ principal components:

$$\frac{\text{tr}(VarY)}{\text{tr}(VarX)} = \sum_{i=1}^{k} \lambda_i \bigg/ \sum_{i=1}^{p} \lambda_i$$

This equation expresses a cumulative measure of the quota of variability (and therefore of the statistical information) 'reproduced' by the first k components, with respect to the overall variability present in the original data matrix, as measured by the trace of the variance–covariance matrix.

In our context, we use principal components as transformed predictors for Y. As a final result, we obtain for each SME a normalized score $\tilde{\pi}_i$, $0 < \tilde{\pi}_i < 1$, for the operational risk side, which can therefore be merged with the financial score.

In order to integrate both scores, and consequently credit and operational risks, we use the following schema:

1. Starting from quantitative financial data, on the basis of the logistic regression, derive for each SME, $i = 1, \ldots, n$, the corresponding score, π_i.

2. Starting from operational data, on the basis of the selected principal components, derive for each SME, $i = 1, \ldots, n$, the corresponding score, $\tilde{\pi}_i$.

3. Create a precision indicator

$$\delta_i = \frac{\dfrac{1}{Var(\pi_i)}}{\dfrac{1}{Var(\pi_i)} + \dfrac{1}{Var(\tilde{\pi}_i)}}$$

where $Var(\pi_i)$ and $Var(\tilde{\pi}_i)$ are the variance of π_i and $\tilde{\pi}_i$ computed on r bootstrapped samples.

4. Derive a global score π_i^* as a linear combination of π_i and $\tilde{\pi}_i$ weighted by δ_i, $\pi_i^* = \delta_i \pi_i + (1 - \delta_i)\tilde{\pi}_i$.

The approach could also be extended by taking into account nonlinear effects among variables. In particular, further research will consider a probabilistic approach for operational risk modeling as reported in Dalla Valle and Giudici (2008).

8.3   Application

The data set was provided by Tadiran, a telecommunications company which offers enterprises complete converged communications solutions that support voice, data, video and advanced telecom applications (MUSING, 2006). It is represented by a global network of more than 200 distributors and affiliates in 40 countries. Tadiran provided quantitative and qualitative data for a representative subset of SMEs.

The quantitative data at hand was extracted from balance sheets of those SMEs. Table 8.1 presents a subset of the data.

In Table 8.1, the statistical unit is the SME and the relevant financial ratios are: return on equity, non-current assets, net cash from regular operations, equity, current assets, current liabilities, current ratio, equity to balance sheet total, pre-tax profit and net profit.

Table 8.2 shows, for the same SMEs, the operational data. More precisely, for each technical problem (hardware, interface, netcomms, security and software) we collect the associated risk levels. Table 8.2 reports, for each SME and for each technical problem, the related risk (1: low risk, 2: medium risk, 3: high risk).

Note that Table 8.1 and Table 8.2 are linked by the same statistical unit which corresponds to the SMEs that have contracted to Tadiran the maintenance service of their telecommunications equipment. This implies payment of a monthly fee by the SME. For this reason, it is important to assess the financial condition of the SMEs in order to determine that they can actually pay for the service.

Table 8.1 and Table 8.2 show respectively the input data source for logistic regression and principal components analysis. For descriptive purposes, and in order to discover multivariate correlations among the variables, we perform a principal components analysis also on the financial data. The results of such an analysis of the financial data are plotted in Figure 8.1. We remark that, in this case, the first two components account for 70% of the variance.

Table 8.1   Financial XBRL data.

SME	Balance_Sheet_Total	Current_Assets	Non_Current_Assets	Equity
90263	11 794 200	109 21 774	393 717	627 750
91234	47 947	34 813	13 134	27 058
91332	519 440	130 111	389 330	62 568
91375	111 200	94 573	16 627	12 914
91460	160 747	120 877	39 870	111 038
91888	230 3915	1 101 978	1 201 937	591 154
93157	97 970	70 909	27 061	34 268
94098	127 953	75 586	52 367	16 526

Table 8.2   Operational data.

PBX_No	Hardware	Interface	NetComm	Security	Software
90263	1	1	3	1	3
91234	1	1	3	1	3
91332	1	2	3	1	3
91375	2	2	3	1	1
91460	1	1	3	3	3
91888	2	2	3	2	3
93157	1	1	3	3	1
94098	1	1	3	1	1

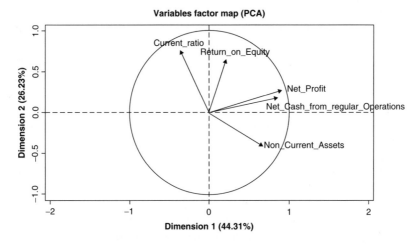

Figure 8.1   Statistical correlations.

Figure 8.1 shows the multivariate correlations among variables, expressed by the cosine of the angle between the corresponding variables. Note that the angle can go from $0°$ to $180°$. Smaller angle cosines between variables show higher positive correlations; if the angle is equal to $90°$ (the cosine is zero) the variables are not correlated and, finally, if the angle is equal to $180°$ (the cosine is -1) a negative correlation is present. In Figure 8.2 we report the scatterplot of the first two principal components (Giudici, 2003). The figure shows a plot of the data in the space of the first two principal components, with the points labeled by the name of the corresponding SME.

In Figure 8.3 we summarize the histogram of each component variance. The figure shows that the first two components dominate. After these explorative and descriptive results, as described in Section 8.3, we derive for each SME the corresponding probability of default, using logistic regression on the original variables. The results are reported in Table 8.3.

In Table 8.3 we present the probability of default coming from logistic regression. Reg1 is based on forward selection, Reg 2 on backward selection and Reg 3

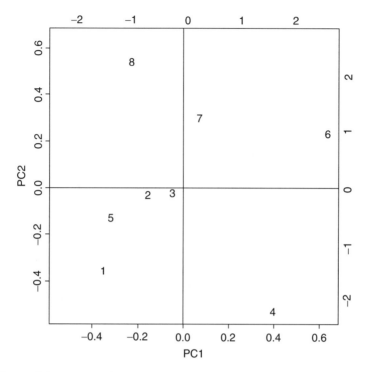

Figure 8.2   Scatterplot of the (scaled) first two principal components.

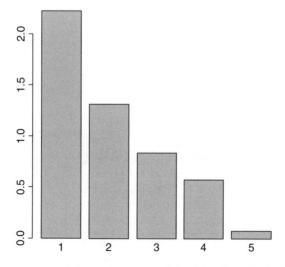

Figure 8.3   Histogram of the variances explained by the principal components.

Table 8.3   Estimated probability of default for the considered SME.

SME	Reg 1	Reg 2	Reg 3
90263	0.590 413 725	0.483 517 201	0.601 052 05
91234	1	1	1
91332	0.198 422 752	0.288 418 151	0.193 413 586
91375	0.311 519 917	0.253 486 66	0.416 534 7
91460	0.585 670 25	0.481 172 502	0.596 798 67
91888	0.593 873 953	0.487 455 978	0.604 682 691
93157	0.606 705 681	0.499 173 616	0.616 373 138
94098	0.598 456 64	0.492 608 605	0.609 354 881

on stepwise selection. Considering the performance indicators (Kenett and Zacks, 1998; Giudici, 2003), we select Reg 3 as the best model,.

We now move to operational data analysis. The data collected on operational risks was summarized in Table 8.2. Starting from this table and applying principal components analysis to the loss data, we can derive similarities among SMEs, labeled by points, as was done for the financial data in Figure 8.2.

As reported in Figure 8.4, SME 13 and 5 are very different in terms of technical problems. The different features from operational risk analysis show

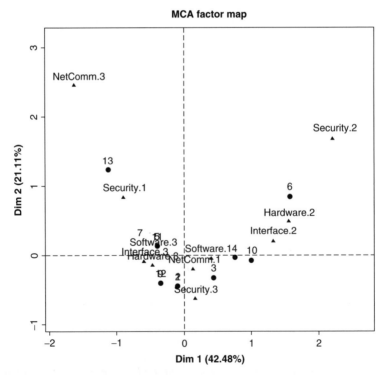

Figure 8.4   The factor map.

that some observations like network communications, none and low-severity-level security are rare cases, while medium-severity cases of network communications and software are very much related. From Figure 8.5, we see that the first two factors cumulatively predict about 65% of the variance.

Finally we compute the precision indicator defined in Section 8.2. The δ_i calculated is reported in Table 8.4.

After the application of the previous procedures, the final score can be estimated on the basis of a linear combination of the financial and operational

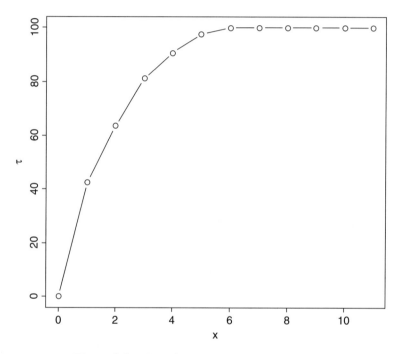

Figure 8.5   Cumulative variance explanation.

Table 8.4   Precision indicator.

SME	Delta
90263	0.784 318 511
91234	1
91332	0.359 319 49
91375	0.767 159 992
91460	0.418 501 195
91888	0.863 035 666
93157	0.758 223 351
94098	0.529 748 722

scores, weighted by the precision indicator. The results of data integration are summarized in Table 8.5, which displays, for each SME, the specific scores.

Table 8.5   Score for each company.

SME	Financial score	Operational score
90263	0.601 052 05	0.453 472 222
91234	1	1
91332	0.193 413 586	0.440 277 778
91375	0.416 534 7	0.247 916 667
91460	0.596 798 67	0.534 722 222
91888	0.604 682 691	0.360 416 667
93157	0.616 373 138	0.417 361 111
94098	0.609 354 881	0.472 222 222

Finally, in Table 8.6 we report the financial, operational and merged integrated scores.

Table 8.6   Financial, operational and merged score for each company.

SME	Financial score	Operational score	Delta	Merged score
90263	0.601 052 05	0.453 472 222	0.784 318 51	0.569 221 813
91234	1	1	1	1
91332	0.193 413 59	0.440 277 778	0.359 319 49	0.351 574 662
91375	0.416 534 7	0.247 916 667	0.767 159 99	0.377 273 676
91460	0.596 798 67	0.534 722 222	0.418 501 2	0.560 701 29
91888	0.604 682 69	0.360 416 667	0.863 035 67	0.571 226 958
93157	0.616 373 14	0.417 361 111	0.758 223 35	0.568 256 677
94098	0.609 354 88	0.472 222 222	0.529 748 72	0.544 868 073

Figure 8.6 plots the densities for the individual scores derived respectively from financial operational data and our merged proposal (see Table 8.6).

In terms of an intuitive score comparison, Figure 8.7 plots, with a radar graph, the financial, operational and merged scores. As we can observe from Figure 8.7, for SME 91332 (labeled 3) the relative financial score is less than the operational score; on the other hand, in SME 91375 (labeled 4) the financial score is greater than the operational score. It is important to note that our proposal allows, in both cases (SME 91332 and SME 91375), a correct measure of the merged score. As pointed out in Section 8.4, this is a correct conservative combination of financial risk and operational risk.

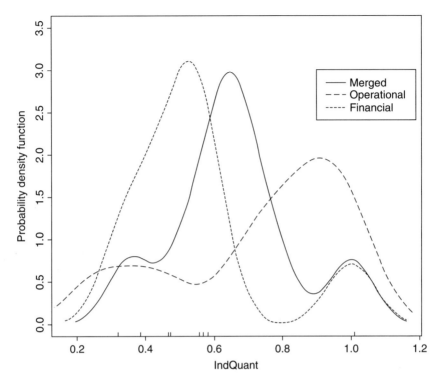

Figure 8.6   Non-parametric density estimation.

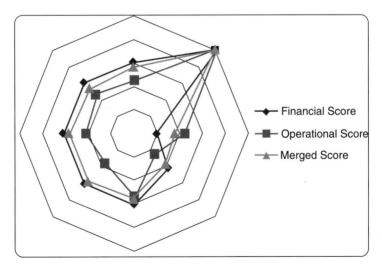

Figure 8.7   Risk comparison.

8.4 Summary

This chapter presents a methodology for measuring, in a combined way, both financial and operational risks. We show how statistical analysis can be successfully used to estimate dependency among operational and financial risks (see also Chapters 7 and 11). Further research is needed to generalize our results to a larger framework. For more information on integrated models, including a novel application of nonlinear principal components analysis, see Figini *et al.* (2010).

References

Alexander, C. (2003) *Operational Risk: Regulation, Analysis and Management*, Financial Times/Prentice Hall, London.

Basel Committee on Banking Supervision (2006) Basel II: International Convergence of Capital Measurement and Capital Standards: A Revised Framework – Comprehensive Version. http://www.bis.org/publ/bcbs128.htm (accessed 21 May 2010).

Cruz, M.G. (2002) *Modelling, Measuring and Hedging Operational Risk*, John Wiley & Sons, Ltd, Chichester.

Dalla Valle, L. and Giudici, P. (2008) A Bayesian approach to estimate the marginal loss distributions in operational risk management, *Computational Statistics and Data Analysis*, 52, 6, pp. 3107–3127.

Figini, S., Giudici, P., Sanyal, A and Uberti, P. (2007) A statistical method to optimize the combination of internal and external data in operational risk measurement, *Journal of Operational Risk*, 2, 4, pp. 69–78.

Figini, S., Kenett, R.S. and Salini, S. (2010) Integrating operational and financial risk assessments, *Quality and Reliability Engineering International*, http://services.bepress .com/unimi/statistics/art48 (6 March 2010).

Fuchs, C. and Kenett, R.S. (1998) *Multivariate Quality Control: Theory and Applications*, Chapman and Hall, London.

Giudici, P. (2003) *Applied Data Mining*, John Wiley & Sons, Ltd, Chichester.

Kenett, R.S. and Zacks, S. (1998) *Modern Industrial Statistics: Design and Control of Quality and Reliability*, Duxbury Press, San Francisco.

MUSING (2006) IST-FP6 27097, http://www.musing.eu (accessed 21 May 2010).

XBRL, http://www.xbrl.org/Home/ (accessed 8 March 2010).

9

Measures of association applied to operational risks

Ron S. Kenett and Silvia Salini

9.1 Introduction

Association rules are one of the most popular unsupervised data mining methods (Agrawal *et al.*, 1993; Borgelt *et al.*, 2004; Kenett and Salini, 2008a, 2008b; Roever *et al.*, 2008; Tan *et al.*, 2004). They were developed in the field of computer science and typically used in applications such as market basket analysis, to measure the association between products purchased by consumers, or in web clickstream analysis, to measure the association between the pages seen by a visitor to a site. Sequence rules algorithms are employed to analyse also the sequence of pages seen by a visitor.

Association rules belong to the category of local models, that is methods that deal with selected parts of the data set in the form of subsets of variables or subsets of observations, rather than being applied to the whole database. This element constitutes both the strength and the weak point of the approach. The strength is that, in being local, they do not require a large effort from a computational point of view. On the other hand, the locality itself means that a generalization of the results cannot be allowed – not all the possible relations are evaluated at the same time.

Mining frequent *itemsets* and association rules is a popular and well-researched method for discovering interesting relations between variables in large databases. Piatetsky-Shapiro (1991) describes analysing and presenting

Operational Risk Management: A Practical Approach to Intelligent Data Analysis   Edited by Ron S. Kenett and Yossi Raanan   © 2011 John Wiley & Sons, Ltd

meaningful rules discovered in databases using different measures of interest. The structure of the data to be analysed is typically referred to as transactional in a sense explained below.

Let $I = \{i1, i2, \ldots, in\}$ be a set of n binary attributes called 'items'. Let $T = \{t1, t2, \ldots, tm\}$ be a set of transactions called the database. Each transaction in T has a unique transaction ID and contains a subset of the items in I. Note that each individual can appear more than once in the data set. In market basket analysis, a transaction means a single visit to the supermarket, for which the list of products bought is recorded. In web clickstream analysis, a transaction means a web session, for which the list of all visited web pages is recorded (for more on clickstream analysis and web usability see Kenett et al., 2009). From this very topic-specific structure, a common data matrix can be easily derived, a different transaction (client) for each row and a product (page viewed) for each column. The internal cells are filled with 0 or 1 according to the presence or absence of the product (page).

A rule is defined as an implication of the form $X \Rightarrow Y$ where $X, Y \in I$ and $X \cap Y = \phi$. The sets of items (for short *itemsets*) X and Y are called antecedent (left hand side or LHS) and consequent (right hand side or RHS) of the rule. In an *itemset*, each variable is binary, taking two possible values only: '1' if a specific condition is true, '0' otherwise.

Each association rule describes a particular local pattern, based on a restricted set of binary variables, and represents relationships between variables which are binary by nature. In general, however, this does not have to be the case and continuous rules are also possible. In the continuous case, the elements of the rules can be intervals on the real line that are conventionally assigned a value of TRUE $= 1$ and FALSE $= 0$. For example, a rule of this kind can be $X > 0 \Rightarrow Y > 100$.

Once obtained, the list of association rules extractable from a given data set is compared in order to evaluate their importance level. The measures commonly used to assess the strength of an association rule are the indexes of *support, confidence* and *lift*:

- The **support** for a rule $A \Rightarrow B$ is obtained by dividing the number of transactions which satisfy the rule, $N\{A \Rightarrow B\}$, by the total number of transactions, N

$$support\{A \Rightarrow B\} = N\{A \Rightarrow B\}/N$$

The support is therefore the frequency of events for which both the LHS and RHS of the rule hold true. The higher the support, the stronger the information that both types of events occur together.

- The **confidence** of the rule $A \Rightarrow B$ is obtained by dividing the number of transactions which satisfy the rule $N\{A \Rightarrow B\}$ by the number of transactions which contain the body of the rule A

$$confidence\{A \Rightarrow B\} = N\{A \Rightarrow B\}/N\{A\}$$

The confidence is the conditional probability of the RHS holding true given that the LHS holds true. A high confidence that the LHS event leads to the RHS event implies causation or statistical dependence.

- The **lift** of the rule $A \Rightarrow B$ is the deviation of the support of the whole rule from the support expected under independence given the supports of the LHS (A) and the RHS (B)

$$lift\{A \Rightarrow B\} = confidence\{A \Rightarrow B\}/support\{B\}$$

$$= support\{A \Rightarrow B\}/support\{A\}support\{B\}$$

Lift is an indication of the effect that knowledge that LHS holds true has on the probability of the RHS holding true. Hence lift is a value that gives us information about the increase in probability of the 'then' (consequent RHS) given the 'if' (antecedent LHS) part:

— When lift is exactly 1: No effect (LHS and RHS independent). No relationship between events.

— For lift greater than 1: Positive effect (given that the LHS holds true, it is more likely that the RHS holds true). Positive dependence between events.

— If lift is smaller than 1: Negative effect (when the LHS holds true, it is less likely that the RHS holds true). Negative dependence between events.

Relative linkage disequilibrium (RLD) is an association measure motivated by indices used in population genetics to assess stability over time in the genetic composition of populations (Karlin and Kenett, 1977). This same measure has also been suggested as an exploratory analysis methods applied to general 2×2 contingency tables (see Kenett, 1983; Kenett and Zacks, 1998). To define RLD, consider a transactions set with item A on the LHS and item B on the RHS of an association rule. In a specific set of transactions or itemsets, these two events generate four combinations whose frequencies are described in Table 9.1.

There is a natural one-to-one correspondence between the set of all possible 2×2 contingency tables, such as Table 9.1, and points on a simplex (see Figure 9.1). We exploit this graphical representation to map out association rules. The tables that correspond to independence in the occurrence of A and B correspond to a specific surface within the simplex presented in Figure 9.1. By 'independence' we mean that knowledge of marginal frequencies of A and B is sufficient to reconstruct the entire table, that is the items A and B do not interact. For such rules (tables) lift $= 1$ and $D = 0$ (D is defined below).

Let $D = x_1 x_4 - x_2 x_3$, $f = x_1 + x_3$ and $g = x_1 + x_2$, where $f =$ relative frequency of item B and $g =$ relative frequency of item A.

The surface in Figure 9.1 corresponds to contingency tables with $D = 0$ (or $lift = 1$). It can be easily verified that

Table 9.1   The association rules contingency table of A and B.

	B	$\hat{B}$
A	x_1	x_2
$\hat{A}$	x_3	x_4

$$\sum_{i=1}^{4} x_i = 1,\ 0 \le x_i,\ i = 1 \ldots 4.$$

$x_1 =$ the relative frequency of occurrence of both A and B.
$x_2 =$ the relative frequency of transactions where only A occurs.
$x_3 =$ the relative frequency of transactions where only B occurs.
$x_4 =$ the relative frequency of transaction where neither A or B occur.

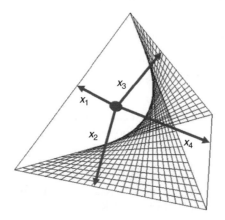

Figure 9.1   The surface of independence (D=0).

$$x_1 = fg + D = support\{A \Rightarrow B\}$$
$$x_2 = (1 - f)g - D$$
$$x_3 = f(1 - g) - D$$
$$x_4 = (1 - f)(1 - g) + D$$

and that

$$confidence\{A \Rightarrow B\} = \frac{x_1}{x_1 + x_2} = \frac{x_1}{g}$$

$$lift\{A \Rightarrow B\} = \frac{x_1}{(x_1 + x_2) \cdot (x_1 + x_3)} = \frac{x_1}{f \cdot g} = 1 + \frac{D}{f \cdot g} \quad (-1 \le D \le 1)$$

The geometric interpretation of D makes it an appealing measure of interaction. As mentioned, the surface on Figure 9.1 represents all association rules with

$D = 0$. However, points closer to the edges of the simplex will have intrinsically smaller values of D.

Let D_M be the distance from the point corresponding to the contingency table on the simplex to the surface $D = 0$ in the direction $(1, -1, -1, 1)$. Then we define RLD $= D/D_M$.

As can be seen geometrically, RLD standardizes D by the maximal distance D_M.

The computation of RLD can be performed through the following algorithm:

If $D > 0$
then
if $x_3 < x_2$
   then $RLD = \dfrac{D}{D + x_3}$
   else $RLD = \dfrac{D}{D + x_2}$

else
if $x_1 < x_4$
   then $RLD = \dfrac{D}{D - x_1}$
   else $RLD = \dfrac{D}{D - x_4}$

Asymptotic properties of RLD are available in Kenett (1983) and RLD can also be used for statistical inference.

9.2   The arules R script library

The **arules** extension package for R (Hahsler *et al.*, 2005, 2008) provides the infrastructure needed to create and manipulate input data sets for the mining algorithms and for analysing the resulting *itemsets* and rules. Since it is common to work with large sets of rules and *itemsets*, the package uses sparse matrix representations to minimize memory usage. The infrastructure provided by the package was also created explicitly to facilitate extensibility, both for interfacing new algorithms and for adding new types of interest measures and associations.

The library **arules** provides the function interestMeasure() which can be used to calculate a broad variety of interest measures for *itemsets* and rules. All measures are calculated using the quality information available from the sets of *itemsets* or rules (i.e. support, confidence, lift) and, if necessary, missing information is obtained from the transactions used to mine the associations. For example, available measures for *itemsets* are:

- All-confidence (Omiecinski, 2003)

- Cross-support ratio (Xiong *et al.*, 2003).

For association rules the following measures are implemented:

- Chi-square measure (Kenett and Zacks, 1998)

- Conviction (Brin *et al.*, 1997)

- Hyper-lift and hyper-confidence (Hahsler *et al.*, 2006)

- Leverage (Piatetsky-Shapiro, 1991)

- Improvement (Bayardo *et al.*, 2000)

- Several measures from Tan (2004) (e.g. cosine, Gini index, ϕ-coefficient, odds ratio)

- RLD (Kenett and Salini, 2008a, 2008b).

As mentioned above, RLD is in the function InterestMeasure(). We use the functions quadplot() and triplot() of the library **klaR** (Roever *et al.*, 2008) to produce the simplex 3D and 2D representation.

9.3   Some examples

9.3.1   Market basket analysis

The first example that we consider is an application to a classical market basket analysis data set. The Groceries data set, available with the arules package, contains 1 month (30 days) of real-world point of sale transaction data from a typical local grocery outlet (Hahsler *et al.*, 2008). The data set contains 9835 transactions and the items are aggregated into 169 categories.

In order to compare the classical measure of association rule with RLD, we plot in Figure 9.2 measures of the 430 rules obtained with the a priori algorithm setting minimum support equal to 0.01 and minimum confidence to 0.1.

The plot shows that RLD, like confidence and lift, is able to identify rules that have similar support. Moreover, for low levels of confidence, the value of RLD is more variable and therefore more informative. The relationship of RLD with lift is interesting. It seems that RLD can differentiate between groups of rules with similar levels of lift.

Table 9.2 displays the first 20 rules sorted by lift. For each rule, the RLD, the odds ratio and the chi-square values are reported. Figure 9.3 shows the value of RLD versus odds ratio and versus chi square for the top 10 rules.

As we expect for the relationship between RLD and odds ratio, the two measures are coherent but still different. The chi-square values appear not to be correlated with RLD so that the information provided by RLD is not redundant with chi square. Moreover, RLD is more intuitive than the odds ratio and chi square since it has a useful graphical interpretation.

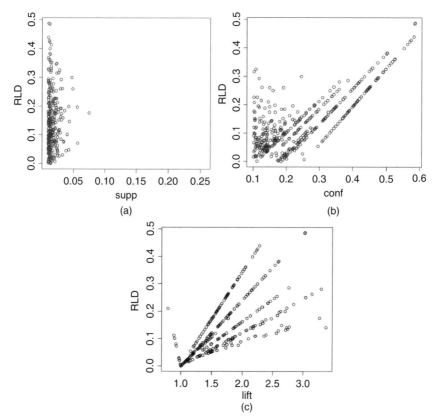

Figure 9.2   Plot of relative linkage disequilibrium versus (a) support, (b) confidence and (c) lift for the 430 rules of Groceries data set.

9.3.2   PBX system risk analysis

In the following example we present an analysis of data collected from private branch exchange (PBX) telecommunication systems discussed in Chapter 5 (see also Cerchiello and Giudici, 2007).

Operational risks, in this context, are typically classified into hardware, software, interface, network and security-related events (see Chapter 3). Assessing operational risks involves merging data from different sources such as system logs, call-centre records, technical service databases and customer complaints (see Chapter 5).

The problem consists of mapping the severity level of problems and the event category (EC) of a PBX under constant monitoring. Seven variables are considered, as shown in Table 9.3. For more details about the data, see Cerchiello and Bonafede (2009).

Table 9.2   First 20 rules for groceries data, sorted by lift.

	lhs		rhs	support	confidence	lift	RLD	oddsRatio	chiSquare
1	{whole milk, yogurt}	=>	{curd}	0.01006609	0.1796733	3.372304	0.1407948	4.565544	184.8700
2	{citrus fruit, other vegetables}	=>	{root vegetables}	0.01037112	0.3591549	3.295045	0.2807587	4.957868	188.4380
3	{other vegetables, yogurt}	=>	{whipped/sour cream}	0.01016777	0.2341920	3.267062	0.1750579	4.449668	177.1536
4	{tropical fruit, other vegetables}	=>	{root vegetables}	0.01230300	0.3427762	3.144780	0.2623764	4.678610	206.0424
5	{root vegetables}	=>	{beef}	0.01738688	0.1595149	3.040367	0.2496032	4.630855	277.3405
6	{beef}	=>	{root vegetables}	0.01738688	0.3313953	3.040367	0.2496032	4.630855	277.3405
7	{citrus fruit, root vegetables}	=>	{other vegetables}	0.01037112	0.5862069	3.029608	0.4869320	6.182676	175.0581
8	{tropical fruit, root vegetables}	=>	{other vegetables}	0.01230300	0.5845411	3.020999	0.4848665	6.194803	207.2034
9	{other vegetables, whole milk}	=>	{root vegetables}	0.02318251	0.3097826	2.842082	0.2253466	4.389810	330.2314
10	{other vegetables, whole milk}	=>	{butter}	0.01148958	0.1535326	2.770630	0.1432227	3.638945	146.3170
11	{whole milk, curd}	=>	{yogurt}	0.01006609	0.3852140	2.761356	0.2855465	4.087797	132.7261
12	{whipped/sour cream}	=>	{curd}	0.01047280	0.1460993	2.742150	0.1345253	3.539378	129.7175
13	{curd}	=>	{whipped/sour cream}	0.01047280	0.1965649	2.742150	0.1345253	3.539378	129.7175
14	{other vegetables, whole milk}	=>	{whipped/sour cream}	0.01464159	0.1956522	2.729417	0.1398891	3.701980	183.7284
15	{other vegetables, yogurt}	=>	{root vegetables}	0.01291307	0.2974239	2.728698	0.2114760	3.791185	163.1868
16	{whole milk, yogurt}	=>	{whipped/sour cream}	0.01087951	0.1941924	2.709053	0.1319696	3.500414	131.6497
17	{other vegetables, yogurt}	=>	{tropical fruit}	0.01230300	0.2833724	2.700550	0.1993601	3.688172	151.3326
18	{root vegetables, other vegetables}	=>	{citrus fruit}	0.01037112	0.2188841	2.644626	0.1484010	3.407110	119.3908
19	{other vegetables, rolls/buns}	=>	{root vegetables}	0.01220132	0.2863962	2.627525	0.1990992	3.568197	141.8140
20	{tropical fruit, whole milk}	=>	{root vegetables}	0.01199797	0.2836538	2.602365	0.1960214	3.513535	136.4357

The data is recoded as a binary incidence matrix by coercing the data set to transactions. The new data sets present 3733 transactions (rows) and 124 items (columns). Figure 9.4 shows the item frequency plot (support) of the item with support bigger than 0.1.

We apply the a priori algorithm to the data, setting minimum support to 0.1 and minimum confidence to 0.8, and obtain 200 rules. The aim of this example is to show the intuitive interpretation of RLD through its useful graphical representation. Figure 9.5 shows the simplex representation of the contingency tables corresponding to these 200 rules. The corners represent tables with relative frequency $(1, 0, 0, 0)$, $(0, 1, 0, 0)$, $(0, 0, 1, 0)$, $(0, 0, 0, 1)$. The dots on the left figure represent all the rules derived from the EC data set and the dots on the right figure correspond to the first 10 rules sorted by RLD.

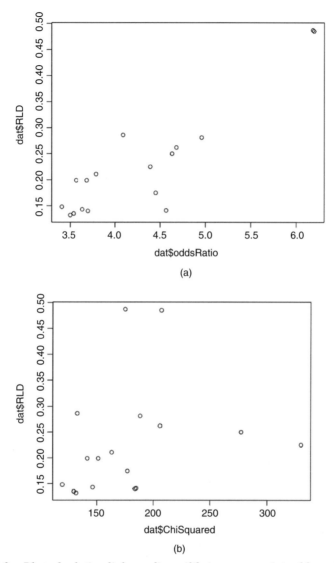

Figure 9.3   Plot of relative linkage disequilibrium versus (a) odds ratio and (b) chi square for the top 10 rules of Groceries data set sorted by RLD.

Figure 9.5 shows that, using a simplex representation, it is possible to have immediately an idea of the rules' structure. In our case, there are four groups of aligned rules. Aligned rules imply that they have the same support.

In order to improve the interpretation, we can try to reduce the dimensionality of the 2 × 2 table. A 2D representation is shown in Figure 9.6. On the bottom left part of the simplex, there are rules with high support, on the bottom right

Table 9.3 Event category data set.

PBX no.	Severity	Customer type	EC2	EC1	ALARM1	ALARM2	ALARM3
90009	2	High tech	SEC08	Security	NO_ALARM	NP	NP
90009	2	High tech	NTC09	Network communications	NO_ALARM	NP	NP
90009	2	High tech	SEC08	Security	NO_ALARM	NP	NP
90009	2	High tech	SEC08	Security	NO_ALARM	NP	NP
90021	2	Municipalities	SEC08	Security	NO_ALARM	NP	NP
90033	2	Transportation	SFW05	Software	PCM TIME SLOT	NP	NP
90033	3	Transportation	INT04	Interface	PCM TIME SLOT	NP	NP
90033	3	Transportation	SEC05	Security	PCM TIME SLOT	NP	NP
90038	2	Municipalities	SFW05	Software	NO_ALARM	NP	NP

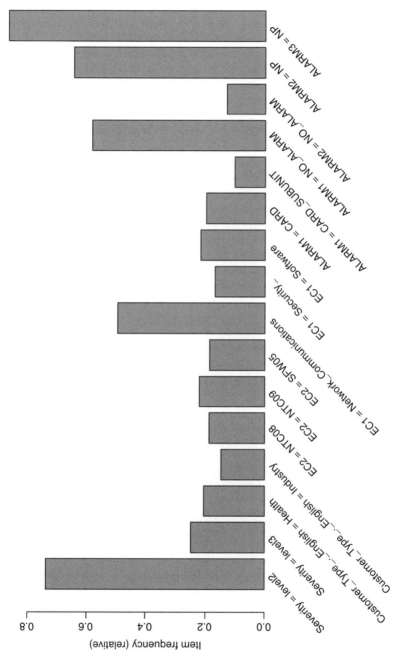

Figure 9.4   Item frequency plot (support>0.1) of EC data set.

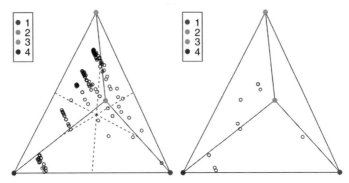

Figure 9.5   The 3D simplex representation for 200 rules of EC data set (left) and for the top 10 rules sorted by RLD (right).

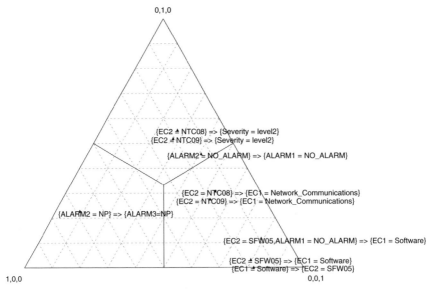

Figure 9.6   The 2D simplex representation for the top 10 rules, sorted by RLD.

there are rules with low support and at the top are the rules with medium support. The table corresponding to the centre point is (0.25, 0.25, 0.25, 0.25).

9.3.3   A bank's operational risk analysis

Operational risk in the banking industry is defined as the risk of loss resulting from inadequate or failed internal processes, people and systems or from external events (Basel Committee on Banking Supervision, 2004). These include:

- Internal fraud
- External fraud

- Employment practices and workplace safety

- Clients, products and business practices

- Damage to physical assets

- Business disruption and system failures

- Execution, delivery and process management

- Includes legal risk.

Operational risks exclude reputational and business/strategic risk.

The rising interest of the banking industry in operational risks is due, among other reasons, to the globalization of the financial markets, the growth of IT applications and the increasing diffusion of sophisticated financial products. The Basel II Capital Accord requires banks to put aside a minimum capital requirement which matches its exposure to credit risk, market risk and operational risk. Specifically, 12% of minimum capital requirement needs to be allocated to operational risks (Basel Committee on Banking Supervision, 2004).

The Basel II Agreement splits operational risk exposures and losses into a series of standardized business units, called *business lines*, and into groups of operational risk losses according to the nature of the underlying operational risk event, called *event types*. In Basel Committee on Banking Supervision (2008), a comprehensive Loss Data Collection Exercise (LDCE) initiated by the Basel II Committee, through the work of its Operational Risk Subgroup of the Accord Implementation Group, is described. The exercise follows other similar exercises sponsored by the Basel Committee and individual member countries over the last five years. The 2008 LDCE is a significant step forward in the Basel Committee's efforts to address Basel II implementation and post-implementation issues more consistently across member jurisdictions. While similar to two previous international LDCEs, which focused on internal loss data, this LDCE is the first international effort to collect information on all four operational risk data elements: (1) internal data, (2) external data, (3) scenario analysis and (4) business environment and internal control factors (BEICFs). The BEICFs are used in an advanced measurement approach (AMA) for calculating operational risk capital charges under Basel II. As an independent contribution to the LDCE we present here the application of RLD to internal operational risk data collected by a large banking institution. Our goal is to demonstrate, with a concrete example, how RLD can be used to assess risks reported in such organizations using textual reports.

We consider a data set of operational risk events with 20 variables, some categorical, some continuous and some textual, with a description of the loss event. Examples of such descriptions are:

- 'Booked on fixed income trade that was in the wrong pat fund code. Have cancelled trade resultant in error of 15000.'

- 'Cash contribution not invested due to incorrect fax number used by client. Not our error but noted due to performance impact on the fund.'

- 'The client sent a disinvestment instruction that was incorrectly processed as an investment. Due to a positive movement in the equity markets the correction of the error led to a gain.'

In the data preparation phase, we discretized the continuous variables (expected and actual values of loss) and, using the library **tm** of R (Feinerer, 2007), we selected the textual description variables, in particular *activity, process* and *risk* type. Then, the data was processed for an association rules analysis.

Following these steps, we obtain a new data set with 2515 transactions and 235 items (the levels of the variables). The a priori algorithm produces 345 575 rules. We modify the default level of support in the arules algorithm of R, and set a very low level of support, 0.01. This is useful in operational risk application, because we expect that the loss events are not so frequent. With such a large number of rules, traditional measures of association typically cannot identify 'interesting' associations – too many rules with too little difference between them. Moreover, with traditional measures of association, it is often difficult to explore and cluster rules in an association rules analysis. RLD and its complementary simplex representation help us in tackling this problem.

For each rule, we calculate RLD and sort the rules accordingly. Figure 9.7 shows the first 200 rules with the highest level of RLD.

We compare the top 200 rules derived from sorting association rules by support, confidence and lift with RLD (see Figure 9.8). RLD clearly provides the highest resolution and interesting spread.

We proceed with an automatic clustering of the rules. This is applied here to the first 200 rules sorted by RLD, but can also be done for other rules.

The hierarchical cluster analysis is applied to the elements in the association rules contingency table on the numbers that we use in the calculation of RLD. Figure 9.9 shows the cluster dendrogram with a highlight of 12 clusters of association rules.

Now we produce a simplex representation for each one of the clusters. Figure 9.10 shows these plots. Rules in the same cluster have a similar type of association. All the rules in these plots have a very high level of RLD, near 1, but different values for the other association measures. For example, the rules in the bottom left corner of the clusters 5, 10 and 12 are characterized by very low support and very high lift. On the contrary, rules in clusters 2 and 3 have high support, high confidence and low lift. In cluster 11, there are rules with confidence equal to 1, lift nearer 1 and very low support. This example demonstrates the unique property of RLD, using a real data set. We conclude with a summary and some direction for future work.

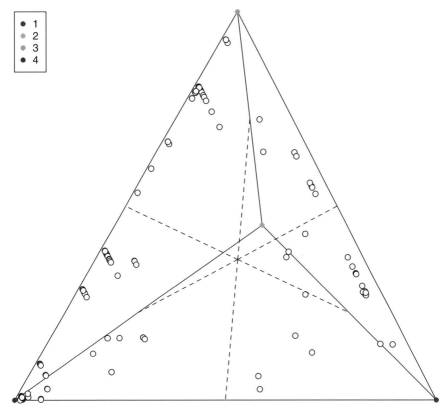

Figure 9.7   Simplex representation of the first 200 rules sorted by RLD for oper-ational risk data set.

9.4   Summary

Relative linkage disequilibrium (RLD) is a useful measure in the context of association rules, especially for its intuitive quantitative and visual interpretation. An inherent advantage to informative graphical displays is that the experience and intuition of the experimenter who collects the data can contribute to the statistician's data analysis. This is an essential component of information quality (InfoQ) discussed in Chapter 1.

The context for applications of RLD ranges over web site logs, customer satisfaction surveys, operational risks data, call-centre records and many other sources of textual data. The first two examples presented in this chapter show that RLD, like confidence and lift, is able to identify rules that have similar

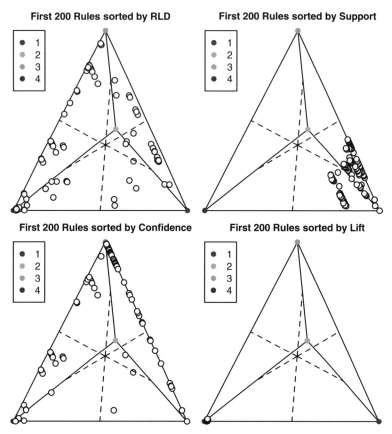

Figure 9.8   Comparison of the first 200 rules sorted by RLD, support, confidence and lift for the operational risk data set.

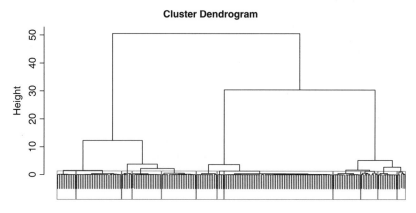

Figure 9.9   Cluster dendrogram for the 200 rules for operational risk data set.

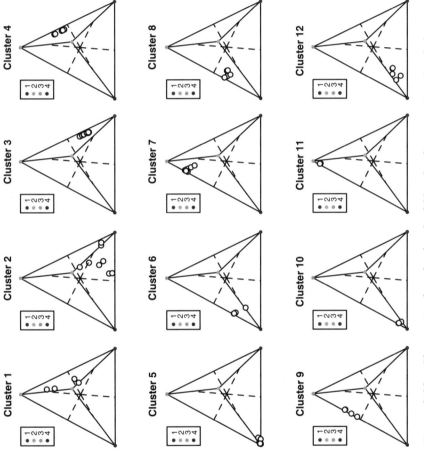

Figure 9.10   Cluster simplex plot for the 200 rules for operational risk data set.

support. Moreover, for low levels of confidence, the value of RLD is more informative. The relationship with lift is interesting; it seems that RLD can differentiate between groups of rules with the same level of lift. RLD is correlated with the odds ratio but differs from the chi-square values. The second example highlights the major advantage of the new measure: it is more intuitive than the odds ratio and chi square and has a useful graphical representation of the rules' structure and allows us to identify groups of rules. The third example shows how RLD can be used to select and cluster association rules.

RLD can contribute to identify rare events in large text files, events called 'black swans' (see Chapter 1, Chapter 14 and Taleb, 2007). Combining RLD with simplex representations can help display item sets with low support, exhibiting significant association patterns. This chapter provides an introduction to RLD with applications to operational risk management. Hopefully it will stimulate more research on association rules and their close relationship with contingency tables.

References

Agrawal, R., Imielienski, T. and Swami, A. (1993) Mining Association Rules between Sets of Items in Large Databases, *Proceedings of the Conference on Management of Data*, pp. 207–216, ACM Press, New York.

Basel Committee on Banking Supervision (2004) Basel II: International Convergence of Capital Measurement and Capital Standards: A Revised Framework, www.bis.org/publ/bcbs107.htm (accessed 21 May 2010).

Basel Committee on Banking Supervision (2008) Operational Risk – 2008 Loss Data Collection Exercise, www.bis.org/publ/bcbs_nl13.htm (accessed 11 January 2010).

Bayardo, R., Agrawal, R. and Gunopulos, D. (2000) Constraint-based Rule Mining in Large, Dense Databases, *Data Mining and Knowledge Discovery*, 4, 2/3, pp. 217–240.

Borgelt, C. (2004) Apriori – Finding Association Rules/Hyperedges with the Apriori Algorithm, Working Group Neural Networks and Fuzzy Systems, Otto-von-Guericke-University of Magdeburg, http://fuzzy.cs.uni- magdeburg.de/~borgelt/apriori.html (accessed 5 March 2010).

Brin, S., Motwani, R., Ullman, J. and Tsur, S. (1997) Dynamic Itemset Counting and Implication Rules for Market Basket Data, *Proceedings of the ACM SIGMOD International Conference on Management of Data*, Tucson, Arizona, pp. 255–264.

Cerchiello, P. and Bonafede, E. (2009) A Proposal to Fuzzify Categorical Variables in Operational Risk Management, in *Data Analysis and Classification*, Springer.

Cerchiello, P. and Giudici, P. (2007) Causal Risk Models: A Proposal Based on Associative Classes, *Proceedings of the 2007 Intermediate Conference 'Risk and Prediction'*, SIS-Società Italiana di Statistica, Venice, pp. 545–546.

Feinerer, I. (2007) *tm: Text Mining Package. R package version 0.3*, http://cran.r-project .org/package=tm (accessed 25 January 2010).

Hahsler, M., Grun, B. and Hornik, K. (2005) arules – A Computational Environment for Mining Association Rules and Frequent Item Sets, *Journal of Statistical Software*, 14, 15, pp. 1–25.

Hahsler, M., Hornik, K. and Reutterer, T. (2006) Implications of Probabilistic Data Modeling for Mining Association Rules, in M. Spiliopoulou *et al.* (Eds), *From Data and Information Analysis to Knowledge Engineering: Studies in Classification, Data Analysis, and Knowledge Organization*, pp. 598–605, Springer-Verlag, Berlin.

Hahsler, M., Grun, B. and Hornik, K. (2008) *The arules Package: Mining Association Rules and Frequent Itemsets, version 0.6-6*, http://cran.r-project.org/web/packages/arules/index.html (accessed 21 May 2010).

Karlin, S. and Kenett, R. (1977) Variable Spatial Selection with Two Stages of Migration and Comparisons Between Different Timings, *Theoretical Population Biology*, 11, pp. 386–409.

Kenett, R.S. (1983) On an Exploratory Analysis of Contingency Tables. *The Statistician*, 32, pp. 395–403.

Kenett, R.S. and Salini, S. (2008a) Relative Linkage Disequilibrium: A New Measure for Association Rules, in P. Perner (Ed.), *Advances in Data Mining: Medial Applications, E-Commerce, Marketing, and Theoretical Aspects*, Lecture Notes in Computer Science, Vol. 5077, Springer-Verlag, Berlin.

Kenett, R.S. and Salini, S. (2008b) Relative Linkage Disequilibrium Applications to Aircraft Accidents and Operational Risks, *Transactions on Machine Learning and Data Mining*, 1, 2, pp. 83–96.

Kenett, R.S. and Zacks, S. (1998) *Modern Industrial Statistics: Design and Control of Quality and Reliability*, Duxbury Press, San Francisco.

Kenett, R.S., Harel, A. and Ruggeri, F. (2009) Controlling the Usability of Web Services, *International Journal of Software Engineering and Knowledge Engineering*, 19, 5, pp. 627–651.

MUSING (2006) IST-FP6 27097, http://www.musing.eu (accessed 21 May 2010).

Omiecinski, E. (2003) Alternative Interest Measures for Mining Associations in Databases, *IEEE Transactions on Knowledge and Data Engineering*, 15, 1, pp. 57–69.

Piatetsky-Shapiro, G. (1991) Discovery, Analysis, and Presentation of Strong Rules, *Knowledge Discovery in Databases*, pp. 229–248.

Roever, C., Raabe, N., Luebke, K., Ligges, U., Szepannek, G. and Zentgraf, M. (2008) *The klaR Package: Classification and visualization, version 0.5-7*, http://cran.r-project.org/web/packages/klaR/index.html (accessed 21 May 2010).

Taleb, N. (2007) *The Black Swan: The impact of the highly improbable*, Random House, New York.

Tan, P.-N., Kumar, V. and Srivastava, J. (2004) Selecting the Right Objective Measure for Association Analysis, *Information Systems*, 29, 4, pp. 293–313.

Xiong, H., Tan, P.-N. and Kumar, V. (2003) Mining Strong Affinity Association Patterns in Data Sets with Skewed Support Distribution, in B. Goethals and M.J. Zaki (Eds), *Proceedings of the IEEE International Conference on Data Mining*, November 19–22, Melbourne, Florida, pp. 387–394.

Part IV

OPERATIONAL RISK APPLICATIONS AND INTEGRATION WITH OTHER DISCIPLINES

10

Operational risk management beyond AMA: new ways to quantify non-recorded losses

Giorgio Aprile, Antonio Pippi and Stefano Visinoni

10.1 Introduction

10.1.1 The near miss and opportunity loss project

This chapter is based on a project conducted in one of the main banks in Italy, Monte dei Paschi di Siena (MPS). MPS, the oldest surviving bank in the world, was founded in 1472, before the voyage of Columbus to America, and has been operating ever since. The project was part of MUSING and concentrated on near misses and opportunity losses within the operational risk management domain (MUSING, 2006). The project developed state-of-the-art methods that were tested on the bank's databases by analysing operational risks that affect information technology (IT) systems. The objective is to better understand the impact of IT failures on the overall process of operational risk management (OpR), not only by looking at the risk events with a bottom line effect in the books, but also by drilling down to consider all the potential losses in terms of missed business opportunities and/or near losses that occur because of IT-related incidents.

Indeed, in current practice of financial institutions, only events which are formally accounted for (i.e. which cause losses that are recorded in the bank's books) are considered in the computation of the operational risk capital requirement.

Operational Risk Management: A Practical Approach to Intelligent Data Analysis   Edited by Ron S. Kenett and Yossi Raanan   © 2011 John Wiley & Sons, Ltd

This is a borderline topic in the Basel II Capital Requirement Directive (Basel Committee on Banking Supervision, 2001, 2006), yet the addressed topics are of paramount importance under the implementation of the Pillar 2 requirements of Basel II, which enlarge the scope of operational risk analyses to include reputation and business risks.

The proposed methodology handles the hidden impact of operational risks by considering:

- **Multiple losses.** This verifies the existence of a possible common IT cause from the analysis of already recorded and apparently uncorrelated losses.

- **Opportunity losses.** This quantifies the missed business opportunities caused by IT failures, in terms of either foregone profits or greater expenses.

- **Near misses.** This tracks near misses with KRIs (Key Risk Indicators); that is, operational risk events not producing a loss.

10.1.2   The 'near miss/opportunity loss' service

The MUSING project developed a 'near miss/opportunity loss' application that can be provided as a web service, together with its main inputs, outputs and the supporting tools (see Figure 10.1).

The detection of near misses and opportunity losses due to IT operational risks is mainly based on the semantic analysis of an IT failure diary, which is recorded by the technicians with free-text descriptions of the incidents. Possible alternatives or supplementary inputs can be records of the calls to IT help desks, or the error logs of IT equipment.

The classification of the events that have been detected as potentially harmful is then driven by a proper representation of the concepts related to IT operational risks. In MUSING, a dedicated ontology was used (see Chapter 3). When classified in a risk category, an incident is also associated with some probability distributions of its business impact. These distributions are derived both from the

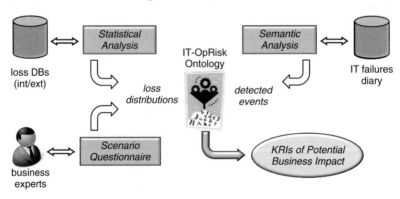

Figure 10.1   Scheme of the 'near miss/opportunity loss' service.

statistical analysis of recorded loss data, which can be either internal to the user or industry wide, and from the execution of scenario questionnaires by business experts, who provide their estimates of the business impacts from the loss events that are associated with each risk category.

Finally, the specific MUSING service returns a KRI that is measured in currency units and quantifies the potential business impact of each recorded IT incident, both as a missed business opportunity and as a near loss. In addition, while the analysis is performed, loss databases are searched and multiple-effect losses are also detected and retrieved.

The outputs of the analysis can be aggregated by the IT component from which the risks arise or by the business unit that is affected by those risks.

10.1.3   Advantage to the user

Through the detection and evaluation of multiple losses, opportunity losses and near misses, the proposed methodology provides an analysis of the impact of IT failures whose main goals are:

- to reconstruct multiple-effect losses, which complies with Basel II requirements;

- to quantify the potential impacts due to reputational and business risks (opportunity losses) and low-level events (near misses), which is indeed a possible extension to the Basel II Advanced Measurement Approach (AMA).

The implemented 'near miss/opportunity loss' service returns early warnings on degraded system performance, and enriches the analysis of the risk profile beyond Basel II compliance. As a consequence, it has an impact both on daily operations and at the regulatory level.

In addition, the methodology can be used to select a mitigation strategy by prioritizing the possible risk-reducing controls and actions that can be taken to prevent a given operational loss. In the common situation where this may occur as the outcome of a chain of IT operational risk events, the service evaluates the effectiveness of intervention at any stage of such a critical path. The cost of the control or action that can be taken against the event at any stage can then be compared with the expected decrease it produces in the KRIs returned by the service. This helps the user select the stage where the risk-reducing intervention is the most efficient. To perform this analysis, the predictive capabilities of the proposed methodology can conveniently be combined with a Bayesian network model of the critical path of risky events.

10.1.4   Outline of the chapter

The chapter is organized as follows. Section 10.2 defines the kind of events dealt with, by giving examples of non-measured losses in a banking context. The methodology is illustrated in Section 10.3, which introduces the reader to how the events need to be classified and what parameters need to be estimated in order to detect 'hidden' losses and quantify their impact. Further details are provided in the case study of Section 10.4, which explains how the 'near miss/opportunity loss' service is applied to MPS's databases. Finally, Section 10.5 summarizes the advantages provided by the methodology, in particular by selecting the best risk-mitigating strategy, and discusses the general scope of its applicability.

10.2   Non-recorded losses in a banking context

10.2.1   Opportunity losses

An opportunity loss is an operational risk event that gives rise to a potential loss because some business opportunity is missed, without any loss being formally accounted for in the bank's books.

As an example, let us consider an IT failure that results in the unavailability of the procedures to close a deal in some branches of the bank. This can occur while customers at the counter are purchasing a specific financial product. Because the IT procedure is unavailable, the deals cannot be closed. As a result of this event, some of the customers who could not buy the financial product will come back and buy it later; but others, who are less determined, will give up the deal and are lost as customers. Therefore, because of the IT failure, the bank will lose a certain amount of profit. The bank undergoes an actual loss (i.e. misses the opportunity of larger profits), yet no loss is recorded in its books. This is consistent with the Basel II definition of operational loss and the loss is not considered in the capital requirements calculations. However, in order to understand better (and mitigate) the impact of operational risks, such missed opportunities should be tracked and evaluated.

The same event can cause both recorded and non-recorded losses. Consider a financial adviser who sets up several appointments with potential customers with all the information about the appointment schedule stored in the electronic calendar of his/her laptop. If a virus attack results in the loss of all the data on the laptop, calendar included, the financial adviser will miss several appointments. Some of these customers can be retrieved later, but some are lost for ever. As a result of the IT event, the pecuniary loss associated with the system recovery of the adviser's laptop will be written in the books. Yet, the major loss the bank suffers is that it misses the opportunity to sign contracts with new customers; this larger loss (i.e. the missed opportunity of larger profits) is not recorded in the books.

In both these examples, the opportunity loss arises from an IT event, and what is missed is the opportunity of otherwise foregone profits. A missed business opportunity may also be caused by a non-IT operational event (e.g. malpractice or fraud by employees). Such events may cause a missed opportunity of carrying out the same business at lower expense. What is important is that missed business opportunities, in terms of either lower profits or greater expenses, do not result in any loss formally accounted for in the books. According to the Basel II definition of operational loss, they make no contribution to the computation of the operational risk capital requirement.

Nonetheless, opportunity losses can help to understand better the impact of IT failures on the overall OpR process. Usually, IT failures are monitored only from a technical viewpoint. No further analysis by 'business-oriented' functions is performed, since such failures do not directly imply accounting losses and are not envisaged as missed business opportunities. Yet, a dedicated processing of the valuable information of opportunity losses can significantly improve the performance of the services rendered as well as mitigation of the existing operational risks.

10.2.2   Near misses

A near loss (or near miss) is an operational risk event that does not produce a loss, neither a pecuniary loss formally accounted for in the books, nor a potential loss, in terms of a missed business opportunity. Again, in case of a near loss, no loss is formally written in the books. For near misses and incidents in industries such as oil and gas, health care and airlines, see Chapter 14.

A near loss can be thought of as an event that could have resulted in a pecuniary loss, if it had occurred in a 'complete' fashion. Accordingly, the event was *near* to producing a *real* loss, but did not develop to the level of its full realization, that is to the stage where it would have produced an actual loss. Likewise, we can think of a near loss as an event that produces a real loss only when it happens together with some other events.

As an example, consider an IT system with several protection levels, for example an IT platform for home banking services with four protection levels (user's ID, a couple of passwords and a public-key certificate). Assume we have observed a trend in violations of the system by hackers' attacks, as pictured in Figure 10.2.

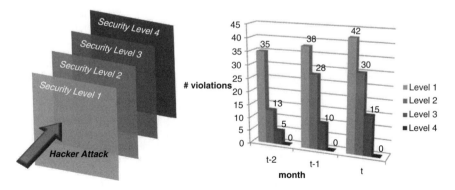

Figure 10.2   Violations of IT protection levels as near misses.

Hackers' attacks do not provoke any pecuniary loss, as long as Level 4 is not violated. Yet, the violations of Levels 1, 2 and 3 are symptoms of an increasingly risky profile, though without any actual loss. Thus, the trend in the number of near loss events provides a very valuable indicator for the identification and prevention at an early stage of degraded IT system performance.

In the example, the near loss is related to the violation of some security levels, with a last security level that is not violated. This prevents the attack from giving rise to an actual loss. This is a common property: the concept of near miss can be associated to that of a loss that would actually have been suffered only if all the steps of a given *path* had been gone through; or, in an alternative view, only if all the required events of a given set happened *together*. This is particularly apparent for the events that affect IT systems, where the definition of such security levels, or simultaneously necessary IT events, rests on a clear technological ground.

The above property turns out to be essential in order to identify, and *measure*, the risk that has been run when a near loss event is detected. In order to derive an indicator of the risk profile from the information about the recurrence of near losses, a level of criticality for the near loss can be derived from the relative number of the necessary IT events that actually occurred, for example the security levels that were violated. This provides a measure of how near the loss was to its actual realization.

Near losses can be tracked with KRIs: they can be the basis for the definition and reporting of a set of appropriate indicators that ensure a more thorough risk profile monitoring. A comprehensive monitoring of the quasi-critical and low-level events recorded by the IT Department provides advance warning of an increased probability of a significant event. This can be achieved, for example, by lowering the 'reporting threshold', as shown in Figure 10.3.

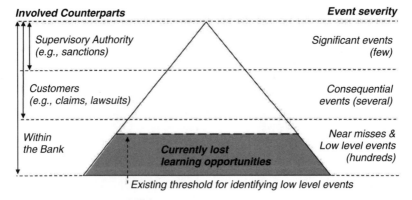

Figure 10.3   Lowering the reporting threshold.

10.2.3   Multiple losses

A multiple-effect loss is an operational risk event that produces pecuniary losses in several business units.

As an example, consider a major crash in the central IT systems of a bank. Because of the crash, many procedures, serving different business units, are unavailable for hours. This results in significant disruption for the bank's customers who cannot use its services (at the counters, online, or at ATMs), therefore possibly provoking many claims and lawsuits against it.

Each business unit or department that suffers a pecuniary loss from the IT downtime records it as a separate event; no evidence is given that it is only a sample of a wider set of linked losses. Therefore, all these operational losses are formally accounted for in the books as separate and uncorrelated events. When the loss database is analysed to evaluate the bank's operational risk profile, it seems that several loss events with limited severity have occurred, instead of a single, extreme IT event. This would significantly change the capital requirement calculated by an internal AMA statistical model. The fact that it is not recorded also prevents risk managers from giving the right priorities to risk-reducing actions.

IT events often result in multiple losses because the same IT system or procedure generally serves more than one business unit, or because of cascading effects in IT systems. Therefore, identifying a common IT cause of already recorded and apparently uncorrelated losses may have a significant impact on the estimation of the operational risk capital requirement. This is also required under the provisions of the supervisory authority. Most of all, tracking such common and underlying causes is of great importance for selecting the best mitigation actions.

Note the difference in the very nature of multiple losses when compared with that of opportunity losses and near misses: *multiple losses* are true pecuniary losses, actually recorded in the books; that is, they are operational losses under the Basel II definition. Yet, it is not trivial to reconnect such multiple losses and write them with the proper value. As for *opportunity losses* and near misses, an intelligent analysis of recorded data is usually required to disclose what actually happened.

10.3   Methodology

10.3.1   Measure the non-measured

Both missed business opportunities and near losses have the characteristic feature that they are not formally accounted for in the books, so they do not contribute to the operational risk capital requirement. Their inclusion can still ensure a better performance of OpR and the mitigation system, also in the view of Pillar 2. Accordingly, the ultimate outcome that is expected from the analysis of opportunity losses and near misses is to 'measure the non-measured': that is, to quantify the pecuniary impact (*in currency units*) of such events. This is intuitive in case of an opportunity loss, and corresponds to providing an estimate for the value (e.g. in euros) of the missed opportunity, in terms of either lower profits or greater expenses.

In case of a near loss, not only has a KRI to be drawn from the recurrence of the event, but that KRI should also be expressed as the amount (e.g. in euros) of the potential damage that the near loss was close to leading to. This means that a proper pecuniary value, associated to a near miss event, should take into account:

- both the *danger level* of the near miss (i.e. the potential loss that it would have caused if completely realized, that is if all the necessary IT events had occurred together – for example, when all the security levels are violated);

- and the *approach level* of the near miss to a realized loss (i.e. a measure of how close it was to producing a true loss – for example, how many of the necessary IT events actually occurred, compared with their total number).

Such a pecuniary value associated with a near miss event does not have to be interpreted in a strict economic sense, since *it measures a loss that did not occur*. It is simply an indicator of the risk profile. Expressing it in currency units may have a much stronger impact on the company management (millions of euros mean much more than any 'red alert', especially to bank executives).

10.3.2   IT events vs. operational loss classes

The need to deal with both *actual* and *near* loss events leads us to draw the following distinction. An *IT event* is any non-standard event in the operation of the IT systems (any problem, failure, attack, etc.) which prevents the system from working properly. A *class* of IT events can be defined by identifying and collecting all the possible events that share the same IT features: they arise from the same *risk factor* and affect the same component of the IT systems (e.g. a hardware break in the operations room; a software crash on terminals; a virus attack on sensitive data repositories; a server failure in the communication network; etc.).

On the other hand, an operational *loss event* is any economic damage suffered by the bank because an operational event prevents it from carrying out its business. Here, we consider economic damage that may be both losses recorded in the books and missed business opportunities. A *class* of loss events is a collection of all possible events that affect the same business unit and have the same kind of *impact on business*. If a class is correctly defined, this homogeneity is essential in order to describe the potential economic damage associated with the events in the class; dealing with similar events that affect a specific business unit makes building such a description easier.

An IT event may result in a loss event; in that case, a cause and effect link can be established between the former and the latter (Kenett, 2007). The previous relation is a many-to-many one: several, different IT events may determine the same loss event (think of the operations of a business unit, usually relying on several IT components); moreover, an IT event may give rise to more than one loss event, as in multiple-effect losses.

Keeping IT and loss events separate also allows near misses to be described. In case of a near miss, one or more IT events occur that are not *sufficient* to cause a loss event, which does not actually occur. Some other concurrent IT events would be needed, but they did not happen. Therefore, in order to detect near misses, for each loss class, it is required to identify sets of concurrently necessary IT events that, when they occur together, result in a loss event; individually, those IT events are the ones that may result in near misses.

Finally, a complete description of the links between the failures of the IT systems and their economic impacts on the business units can be provided by identifying:

- set of business-oriented classes for the loss events (*possible effects*, homogeneous in terms of the business impact);

- set of system-oriented classes for the IT events (*possible causes*, related to the risk factors);

- *mapping* of cause and effect links between the two sets (involving both single IT events and sets of concurrently necessary IT events).

The previous sets of classes and the mapping between the two can be derived:

- either through a *self-assessment* process with business and IT experts (i.e. on the basis of a priori knowledge);

- or through a semantic analysis (*clustering*) of the relevant databases.

The two techniques can be used to refine each other. By assuming matched IT and loss events, through the mapping, we assume that each class of losses is associated with all the possible IT incidents that can produce them.

The approach we take is a *deterministic* mapping of the classes of IT incidents on the classes of losses. This means that when an IT event of one class occurs, only two outcomes are given for each class of losses, depending on whether the former is mapped on the latter (in which case we assume that the IT event *certainly* causes a loss of that class, apart from the amount of the latter) or not (no loss occurs). Yet, a significant improvement of the model can be achieved by introducing a *probabilistic* description of cause and effect links. Thus, having fixed a class of losses and a class of IT incidents, the mapping is represented by the *probability distribution* of a loss *conditional* upon the occurrence of an IT incident. The deterministic approach can then be seen as a special case of the more general probabilistic mapping. The introduction of the probabilistic mapping significantly enriches the description of the actual links that are observed between the occurrence of IT events and losses. The only modification needed in the methodology presented in this chapter is to perform a *convolution product* between the conditional probability distribution of the class of losses, given the class of IT events, and each of the probability distributions for the amount of loss associated to that class (both actual, recorded losses and estimated opportunity losses). The risk measures that are described in Sections 10.3.3, 10.3.4 and 10.4.4 are then taken from the above distributions determined through convolution, instead of the loss distributions themselves. See also Section 10.3.4 for the implicit inclusion of the loss approach level. The main drawback of the probabilistic mapping is the (usually) limited amount of recorded data available in order to derive the conditional probability distributions. For the sake of simplicity in the presentation of the methodology, we apply the simpler deterministic mapping.

10.3.3 Quantification of opportunity losses: likelihood estimates

Since loss classes are homogeneous as to the potential economic damage, each of them can be associated to an estimate of the impact of missed opportunities; this is not an exact value, but a *loss distribution*. Such distributions are determined by means of some *scenario analyses*, where business experts provide their estimates of the potential opportunity loss.

This allows opportunity losses to be quantified as follows. Each incident recorded by the IT Department is assigned to one (or more) class(es) of IT events, by performing annotation and *information extraction* from the analysed records (see Chapter 4). In doing this, it is not necessary to pick up a single class. A better description of the potential impact of the incident is achieved if it is associated, mainly by semantic analysis, to several classes, each with an estimated *likelihood* (see Chapter 3). The likelihood estimate is a value in the [0, 1] range that measures the reliability in stating that the incident under analysis is actually a sample of that class of IT events. Therefore, this association also identifies the possible kind of business opportunities that might have been missed because of the incident. Those are the ones that are linked to the selected class(es) of IT events via the mapping described in Section 10.3.2.

Finally, through the mapping and the likelihood estimates, a risk measure is taken from the loss probability distribution(s) of the selected class(es), which quantifies the pecuniary value of the opportunity loss.

It is recommended to have each class of losses pertain to a specific business unit. This homogeneity simplifies the probabilistic description of the potential loss. The probability distributions of the missed opportunities are derived from interviews with business people, who provide their estimates only for the potential impact of the event on the business unit they deal with.

Yet, the proposed methodology is able to describe the overall impact of an IT event that gives rise to several opportunity losses in different business units or departments. Such 'multiple opportunity losses' are taken into account by the model as the recorded failure is assigned to *a class of IT incidents that is mapped on more than one class of loss events*. It is also worth focusing on the dependence among the likelihood estimates that link a given failure in the records to the different classes of IT events. Two choices are possible: either the sum of the likelihood estimates over all the classes of IT events has to be 1, or they can vary independently (e.g. with several ones, or all zeros, etc.). In principle, both solutions work. A third option is to apply multivariate dependencies in the form of copulae.

10.3.4   Quantification of near misses: loss approach level

In order to estimate the *danger level* of a near miss (i.e. to quantify the potential loss it could have produced), each class of loss events is associated to a distribution describing the pecuniary loss that is suffered when an actual loss occurs, that is when all the necessary IT events occur concurrently. In that respect, the considered loss is consistent with the Basel II definition of operational loss, and the corresponding distribution for each class is derived from the time series of the related losses actually accounted for in the books.

In addition, it is advisable to enrich the description of the potential damage of a near miss by including what could have also been its impact in terms of opportunity loss. This can be achieved by taking into account the probability distributions of the missed business opportunities, previously introduced in Section 10.3.3. Therefore, the potential impact of a near miss event is described by using two loss distributions for each class:

- distribution of the potential operational loss, derived from the time series of recorded, pecuniary losses;

- distribution of the potential missed opportunity, derived from the estimates provided by business people performing a scenario.

The two distributions allow one to take into account not only near losses, but also 'near missed opportunities'.

To complete the description of near misses, each IT event that results in a near loss must be associated to a measure of its *loss approach level* (i.e. a measure of how close it was to producing an actual, realized loss). Many choices may be envisaged. Nevertheless, it is suggested that such IT events, which are not sufficient for an actual loss to occur (e.g. the violations of some – but not the last – security levels), be identified through a process of *self-assessment* with the IT people who are familiar with those events, their recurrence and severity (see Chapter 2).

A very simple approach for providing a measure of the loss approach level for an IT event in the [0, 1] range is based on the conditional probability of the actual loss, given the near loss event. The conditional probability depends only on the class of IT events that is selected by the semantic tool for the given incident and on the loss class(es) which the former is mapped on. If the class of IT events is mapped on more than one loss class, each of them is associated to a specific value of conditional probability. This means that the values for the loss approach level can be computed once and for all, for all the possible pairs composed of an IT event and a related loss class. When a class of IT events is picked out for a given IT incident, the associated value of the loss approach level is also retrieved. Loss approach parameters related to different pairs, composed of a class of IT events and a class of loss events, are independent of each other. The above description can be easily extended to deal with IT events that are sufficient to cause a real loss, by simply letting the loss approach level be 1.

Here, we assume that the conditional probability of a loss, given an IT incident which measures the associated loss approach level, is a *point value* in the [0, 1] range. Yet, in order to obtain a more sophisticated and powerful description, we can think of the loss approach level as the *probability distribution* of the class of losses conditional upon the class of IT events.

This approach is fully consistent with the probabilistic mapping between the classes of IT events and losses, which is discussed in Section 10.3.2. In fact, if this kind of mapping is adopted, the conditional probability distributions of the losses given the IT events are already introduced in the mapping, which therefore *includes* a description of the associated loss approach levels. In such a case, the risk measures taken from the distributions determined through convolution (of the conditional and loss distributions) are already – implicitly – weighted by the loss approach level of the IT event with respect to the class of losses. Again, for the sake of simplicity in the exposition, deterministic mapping and point-value fulfilment parameters are assumed in the following.

Finally, the proposed approach for detecting and quantifying both actual and near losses can be briefly summarized as follows:

- If it is detected that an IT incident in the diary 'went through' and reached its maximum severity, therefore producing actual pecuniary losses (i.e. if it is *likely* that the IT failure comes from a class of IT events that are sufficient to cause losses in one or more classes), these are searched for in the database of recorded losses.

- Otherwise, if it is determined that the incident resulted only in a near miss event:
  — for each class with non-zero likelihood, one *risk measure* (e.g. VaR) is drawn from the probability distribution of the operational pecuniary loss and
  — another from the probability distribution of the missed business opportunity;
  — the two samples are then added up
  — and weighted (multiplied) by the product of the values for the likelihood and the loss approach parameters, estimated for that class.

By performing the previous computation over all the loss classes, a set of class-related potential impacts is determined. Finally, on adding up, the desired KRI is obtained, in the form of a *fictitious but meaningful pecuniary value*, which quantifies the potential damage the incident could have had (in terms of both operational losses and missed opportunities), as well as how close the incident was to causing that effect.

To a rough approximation, the loss approach level can be estimated as a *relative recurrence* (how many times an actual loss is observed when that – not sufficient – IT event occurs). That is, as the ratio between:

- the number of times when both the IT event and an actual loss of the associated class are observed (because all the required IT events concurrently happen) and

- the total number of times when that IT event is observed.

To demonstrate all the previously introduced concepts, let us go back to the example of a home banking platform with four security levels. The violations of each of the four protection steps are the possible classes of IT events; misuse by hackers, after the full theft of the user's identity, is the associated class of losses. An actual loss only occurs when all the security devices are violated, that is when the four IT events happen concurrently, otherwise they can be viewed as simple near misses. When the attack is completely successful, a pecuniary loss is expected, whose probability distribution can be derived from the analysis of the historical series of the loss data that was observed and recorded in such cases (Figure 10.4, left). In addition, further economic damage can be expected from the possibly missed business opportunities. The amount of the associated theoretical (not recorded) loss can be described by means of a probability distribution obtained from business experts' estimates (Figure 10.4, right).

Let us assume that any security level can be violated only after breaking all the previous ones: this means that we have identified a *critical path* of IT events, which ends with a real loss in the last stage if the hacker somehow manages to seize the certificate. We assume that the conditional probability of the loss event (actual misuse by hackers), given each of the IT events, that is the breach of each security level, equals the relative recurrence of the two events. For the occurrences in the table, Figure 10.5 plots the value of the loss approach level of each IT event in the considered critical path.

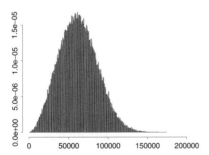 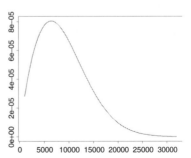

Figure 10.4   Left: empirical distribution of recorded pecuniary losses. Right: parametric distribution of estimated business opportunities.

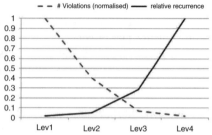

Security level	# attacks tha violated the i-th level	loss approach level (approx as relative recurrence)
Lev1	500	0.0200
Lev2	200	0.0500
Lev3	35	0.2857
Lev4	10	1.0000

Figure 10.5   Relative recurrences of IT events in a critical path, with the associated loss approach levels.

10.3.5   Reconnection of multiple losses

Multiple losses are true pecuniary losses, therefore recorded in the loss databases; here, the goal is to reconnect separate losses originating from a common hidden cause. The trail that links such losses can be found by grabbing keywords from the records in the bank's databases.

The loss databases are analysed, by extracting keywords and other relevant information that may be of use in linking each record to another, and to those of the IT failures stored in the IT Department's diary. Such information may include dates, name of the branch, name of the involved customer, etc. This information is then compared with that relevant to the reported IT failures. For each pair IT incident-recorded loss, a possible cause and effect link is estimated, in the [0, 1] range.

When a reliable link is detected (according to a specified *threshold* probability), the IT incident is associated to that true loss, actually recorded in the books. When more than one recorded loss is found to be linked to the same IT failure, which is therefore identified as the common cause of a *multiple-effect loss*, the IT incident is associated to the sum of all the linked operational losses.

10.4   Performing the analysis: a case study

10.4.1   Data availability: source databases

The proposed methodology has to be applied to input data that allows for information extraction, mainly via semantic analysis, in order to evaluate the likelihood of the link between an IT incident and a recorded loss. Therefore, the methodology best fits source databases that include fields with free-text descriptions of the events (IT failures or operational losses). In the following, a brief overview is provided for MPS's databases on which the research activity was developed and tested.

➢ *Call-centre and technicians' logs and records*

The methodology was tested on MPS's *ICT failure diary* (*Diario Incidenti*), a Web-based diary on the bank's intranet which collects the technicians' records of the reported IT incidents. Each record has some *free-text descriptions* of the IT incident and of the technician's intervention, as well as:

- incident opening time and date, signalling operator ID;

- incident closure time and date, solving operator ID.

It is both a batch and an online input. As a *batch* input, the records in the diary are analysed (mainly semantically) in order to derive rules that are used to detect risky events in any new record, and to classify each of them in a risk category of an IT operational risk dedicated ontology. Once such rules are defined, new records in the IT failure diary can also be supplied as *online* input.

It is worth noting that the diary somehow has a 'service-oriented' structure: several fields in the records aim at providing information on the business unit that has (or may have) been affected by the reported incident. In addition, the organizational model of MPS's IT service company is such that each IT department (the one which took charge of the incident is recorded in the diary) generally corresponds to one of the Bank's high-level business units, served by that department. Therefore, the signalling employee may be a front-office operator, whose business department is also specified; otherwise, even if he/she is an IT operator, his/her department name is still useful to retrieve the business unit that is involved in the incident. This also applies to the operator solving the problem.

As an alternative batch or online input, the error logs of IT equipment can be used.

> *Customer claims and lawsuits records*

The *Claims DB* (*DB Reclami*) is a structured source where customer claims are recorded by the Audit Department. The *Lawsuits DB* (*DB Cause Legali*) relies on a Web-based application where lawsuits are recorded by the Legal Department. It also includes fields with specific OpR-oriented information. Both sources contain:

- *free-text remarks* about the matter of the claim/lawsuit;

- the amount of the economic loss suffered by the bank.

They are the *batch* input of a statistical analysis that returns a probability distribution for the business impact (in terms of actual recorded losses) from the loss events that are associated to each risk category. In order to augment the loss distributions, such input information can also be supplemented by *industry-wide* loss data; for example, that provided for the Italian market by the interbank DIPO Consortium (an Italian database on operational losses in banks).

> *Scenario questionnaires*

For each risk category (i.e. for each class of loss events), specific scenarios are used to collect business experts' estimates of the potential opportunity loss. The answers to the questionnaires are the *batch* input of a statistical analysis that returns a probability distribution for the business impact (in terms of missed business opportunities) from the loss events that are associated to each risk category.

A proper choice for these probabilities can be a parametric description by means of a two-parameter distribution, such as a Weibull or a lognormal distribution. In this case, a pair of estimates is needed for each loss distribution:

- typical amount (in currency units) for the severity of the opportunity loss caused by one loss event of the considered class (i.e. the amount of opportunity loss that is most often incurred);

- worst case (in currency units) of the above severity.

The former can be linked to the mode, or the median, of the probability distribution, the latter to the VaR at a given high quantile of the distribution (e.g. the 99.9th percentile). Regardless, several different choices for the above probability distributions can be envisaged.

10.4.2   IT OpR ontology

The implementation of the proposed methodology in an application for detecting and quantifying near losses and missed business opportunities requires a classification of the loss events resulting from OpR factors in the IT systems; the logical structure of such a classification must be able to handle all the information that is involved in the process outlined in Section 10.3.

This is achieved by using an OpR dedicated ontology (see Chapter 3). Here, we summarize the main features that allow the methodology to be applied. *For each class of loss events*, the following information is provided:

- A free-text description of the loss event, for example specifying the affected business unit, the banking product, the kind of impact on business, etc.

- An array of classes of IT events that are sufficient to produce a loss of the given class; the occurrence of *any* IT event implies the occurrence of an actual loss.

- An array of sets of IT events that can result in a loss but are not sufficient to cause it; an actual loss occurs only if *all* the IT events that make up *any* set in the array concurrently happen. Otherwise, these IT events only result in near losses.

- A probability distribution for the potential pecuniary impact of an operational loss of that class, derived by *loss data analysis* from the loss databases (see Figure 10.4, left).

- A probability distribution of the potential missed business opportunity to be expected with a loss event of that class, derived from business experts' opinion through *scenario analysis* (see Figure 10.4, right).

In addition, *for each class of IT events* that is linked to the given class of losses (either sufficient or not to produce an actual loss), the following information is also provided:

- free-text description of the IT events comprising the class, which must contain all the possible keywords and other relevant information that can be of use in assigning a reported IT incident to that class (e.g. the risk factor, the affected component of the IT systems, *as well as* the business unit/product served by that IT component, etc.).

- The value of the loss approach level in the [0, 1] range, for the considered pair composed of a class of IT events and a class of loss events. For the sake of consistency in the description, the loss approach level of an IT event that is sufficient to produce a real loss is set to 1.

A simple scheme that sketches the 'concepts' involved in the above IT OpR dedicated ontology and their main attributes is presented in Figure 10.6.

10.4.3   Critical path of IT events: Bayesian networks

As previously explained, the loss approach level takes the meaning of the conditional probability of a *joint occurrence* of all the IT events in a set of concurrently

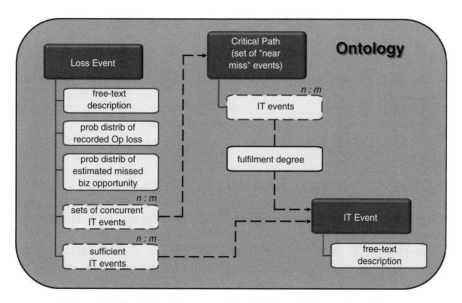

Figure 10.6   Main features of the IT OpR dedicated ontology.

necessary events, given the occurrence of the one under consideration; only when *all* the events in the set happen together/in chain do they cause the realization of an actual loss. The same class of IT events may be part of more than one set of concurrent events, with a different value of the loss approach level in each set. In particular, the same IT event may be both a sufficient cause for one or more loss classes and – at the same time – part of one or more sets of events that *conjointly* lead to some other classes of losses.

When such events can only happen in a chain reaction, i.e. according to some given order, they identify a *critical path* of IT events, whose eventual outcome is an operational loss (see Figure 10.7). Any critical paths of events can be cast in the form of a *Bayesian network*. This helps the modelling and evaluation of the loss approach parameters, each being the probability of a leaf node (i.e. an actual loss) conditional upon the given internal node (i.e. an IT event), for the specific Bayesian network (i.e. the set of concurrent IT events, or the critical path), under analysis. Bayesian networks have been used in a variety of similar applications such as analysing the usability of web sites (Harel *et al.*, 2008) and analysis of customer satisfaction surveys (Kenett and Salini, 2009). For tutorials and reference material on Bayesian networks, see Pearl (1995, 2000), Charniak (1991), Heckerman (1995), Cowell *et al.* (1999), Jensen (2001), Murphy (2001), Cornalba *et al.* (2007) and Ben Gal (2007). For software applications implementing Bayesian networks, see bnlearn (2008), GeNIe (2006) and Huggin (2007).

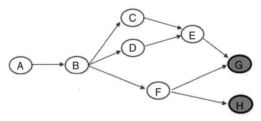

Figure 10.7   Critical paths of IT events: nodes G and H represent real operational losses.

In addition, combining the proposed methodology for the evaluation of near misses with Bayesian modelling further enhances the predictive capabilities of the former. Particularly meaningful is the analysis of the effectiveness of risk-reducing controls and actions on the different nodes of the Bayesian network (i.e. against the different IT events along the critical path), where the cost of intervening is compared with the estimated benefit that follows from the decrease that can be achieved in the conditional probability of a real loss, that is in the loss approach level.

Apart from its implementation in combination with Bayesian networks, the proposed model of near misses is also suitable for sets of events that do not occur in a fixed order, that is when the only significant information is that the loss occurs when all the IT events of a set concurrently happen.

10.4.4   Steps of the analysis

> ➤ *Step 1: estimating likelihoods with respect to classes of IT events*

Given an incident reported in the IT failures diary, the 'near miss/opportunity loss' service analyses the corresponding record (dates, textual description, etc.) and draws out keywords and other relevant information. Such information is interpreted by the service vis-à-vis the domain ontology ('reasoning'), by comparing that information with what is extracted by analysing the textual descriptions of all the classes of IT events (either sufficient or not to produce a loss).

For each class of IT events, a *likelihood parameter* is returned in the [0, 1] range, which measures how far its textual description is ontologically linked to that of the incident reported in the diary. In evaluating likelihoods, the service can also take as an input the textual descriptions of the loss classes (which are linked to the classes of IT events via the mapping). This aims at making the most of the information about the affected business unit that can be retrieved from the IT failures diary.

Therefore, the outcome of Step 1 (see Figure 10.8) is a set of likelihood estimates for the possible classes of IT events that may have occurred (i.e. which the reported failure comes from).

> ➤ *Step 2: processing loss approach levels*

For each class of loss events, it must be assessed *if* – and *to what extent* - the given IT incident has resulted in a near loss. Any class of IT events with non-zero likelihood, which comprises a set of concurrently necessary IT events, provides a *likelihood estimate of a near loss* due to that critical path (i.e. that set). Thus:

- the service retrieves the loss approach level of that class of IT events;

- the estimates for the likelihood of the given incident with respect to the class and for the loss approach level are *multiplied*; and

- then they are *added up* over all the critical paths/sets that are associated to the loss class.

In such a way (see Figure 10.9), the service assigns to each loss class a *combined weight* – depending on both its likelihood and loss approach level – which describes the potential impact of the given incident as a near miss, by taking into account all the possible critical paths and/or joint occurrences of events that could have led to an actual loss.

> ➤ *Step 3: retrieving recorded losses*

The service analyses the records of OpR loss data that are collected in the Claims and Lawsuits DBs, and draws out keywords and other relevant information (dates, branch, textual description, etc., plus the operational pecuniary loss).

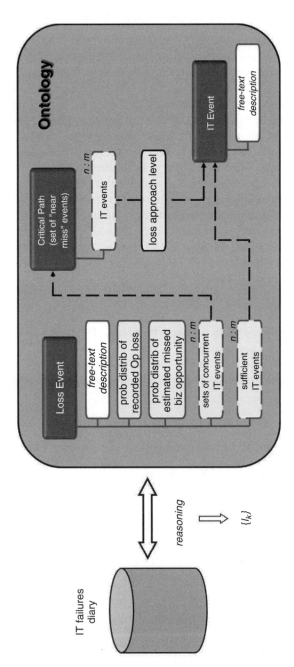

Figure 10.8   Step 1: estimating likelihoods.

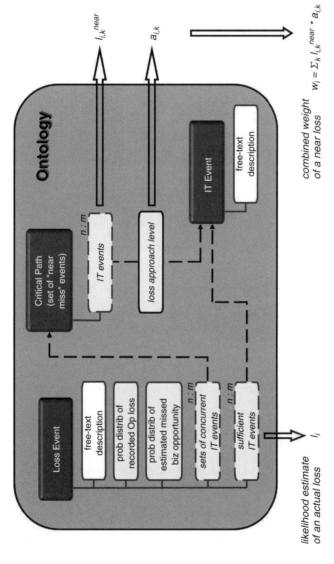

Figure 10.9   Step 2: processing loss approach levels.

If the given IT failure is estimated as likely to have produced a true loss in *at least one class* (i.e. there is at least one non-zero likelihood of an actual loss), the service compares the information drawn out from the IT failures diary with what is obtained by 'reasoning' on *each* record in the Claims and Lawsuits DBs. For each pair, when a reliable cause and effect link is detected (i.e. above a specified *threshold* probability), the recorded pecuniary loss is retrieved.

Finally, all the OpR pecuniary losses – associated to the claims and lawsuits records that are found to be linked to the same IT incident (which may be the cause of a *multiple-effect loss*) – are added up, and the service returns the cumulative recorded loss associated to the IT failure (see Figure 10.10).

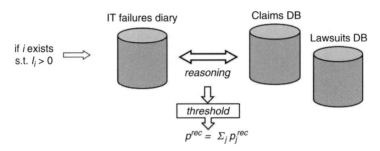

Figure 10.10   Step 3: retrieving recorded losses.

➤ *Step 4: measuring loss distributions*

If the ith loss class is picked out as likely to be linked to the given IT failure (i.e. either the ith likelihood of an actual loss or the combined weight of a near miss is greater than 0), a *risk measure* (e.g. VaR, median, etc.) is drawn from each of the loss distributions for that class (see Figure 10.11), that is:

- both the ith probability distribution of the *operational pecuniary loss* (from loss data analysis)

- and the ith probability distribution of the *missed business opportunity* (from scenario analysis).

The measure that is actually taken from the distributions can be chosen depending on the subsequent analysis to be performed on the loss data.

➤ *Step 5: weighting loss measures*

If the given IT failure is estimated as likely to have gone through and caused actual damage as a loss event of the ith class (i.e. the ith likelihood of an actual loss is greater than 0), the measure of the ith *missed opportunity distribution* is weighted by multiplying by the ith *likelihood* parameter:

$$\text{if } l_i > 0 \quad \Longrightarrow \quad l_i * m_i$$

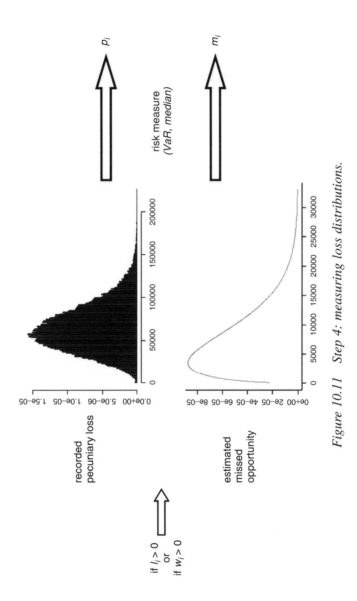

Figure 10.11 Step 4: measuring loss distributions.

Otherwise, if the IT failure is only detected as a *near loss* of the ith class (i.e. the ith combined weight of a near loss is greater than 0), the KRI associated to that class is obtained by adding up the risk measures of *both* the ith loss distributions and by multiplying by the ith *combined weight* of a near miss:

$$\text{if } w_i > 0 \quad \Longrightarrow \quad KRI_i = w_i * (p_i + m_i)$$

➤ *Step 6: updating the IT failures repository*

The values that have been calculated for each ith loss class and the given IT incident reported in the diary, that is

- the estimated opportunity loss (if the ith likelihood of an actual loss is greater than 0) and/or

- the estimated KRI (if the ith combined weight of a near miss is greater than 0),

are respectively added up over all the loss classes. In addition, the same IT incident may have been detected as the actual cause of one or more losses recorded in the DBs; in this case, it has also been associated with the cumulative recorded pecuniary loss. Finally, the corresponding record in the IT failures diary (or in a different repository) is updated with the three pecuniary amounts (see Figure 10.12); each of them may be 0, depending on the nature of the event, level of realization, etc.

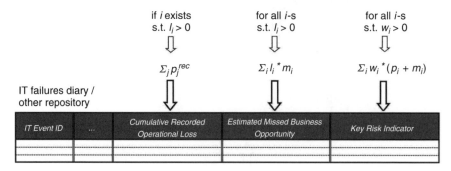

Figure 10.12   Step 6: updating the IT failures repository.

10.4.5   Outputs of the service

All the amounts of loss and all the indicators that are reported by the 'near miss/opportunity loss' service can be quantified and 'grouped':

- *by the IT incident* (or set of incidents) that caused the losses;

- *by the component of the IT system* from which the incidents arose;

- *by the business unit/product* that was affected by the incidents.

The main contents of the reports produced by the service are listed in the following:

- Report on the *actual* business impact, in terms of the amounts of losses recorded in the books. It takes into account all the possible multiple losses caused by a common IT incident. The presence of such losses is identified by also providing:
  - an estimated score of reliability for the link between each loss and the associated IT incident (which may be a multiple-effect event);
  - the criteria used by the system to link the event and the loss (e.g. text analysis, date analysis, location, etc.).

- Report on the *potential* business impact in the form of a KRI that quantifies the risk associated to events that had no bottom line effect in the books. The KRI is measured in currency units and takes into account the potential impact of both opportunity losses and near misses. It provides both:
  - pecuniary amounts of the estimated theoretical/fictitious loss and
  - synthetic indicators that monitor the risk levels associated to near misses and opportunity losses.

10.5 Summary

The novel methodology presented in this chapter provides a 'near miss/ opportunity loss' application for managing 'hidden' and non-measured impacts of operational risks. First of all, it allows the prospective user to estimate the probability and magnitude of such risks, so as to assess his/her tolerance to these kinds of phenomena.

In addition, especially if combined with Bayesian modelling of the possible critical chains of events, the application evaluates the effectiveness of risk-reducing controls and actions, in comparison with their cost. A critical path of low-level events, which can eventually result in an actual significant loss, can be represented in terms of a Bayesian network model. By applying the proposed methodology for the quantification of near misses, a pecuniary KRI can be estimated for each event in the critical path. This helps to select the stage in the critical path where a risk-mitigating action can be most effective and efficient, by comparing the cost of an intervention at that stage with the expected decrease in the KRI it leads to. Therefore, the service provides the user with a consistent prioritization of those actions, in order to define the most suitable mitigation strategy.

Finally, it is also worth noting that the methodology can be easily extended to the more general case of non-IT operational risks, that is to the analysis of non-recorded losses caused by non-IT events. In fact, what is required to make the methodology work is a source database, like MPS's IT failures diary, which includes some textual description of the events that have possibly given rise to some non-recorded losses or to some other kinds of losses we may want to detect. Indeed, this kind of information is often recorded when dealing with IT systems, but if such a source is available the methodology also applies to non-IT operational risks. Customer complaints about the underperformance of financial products and the possible suits for false claims that may follow are significant examples. Other examples from the oil and gas, health care and transportation industries are discussed in Chapter 14.

References

Basel Committee on Banking Supervision (2001) QIS 2 – Operational Risk Loss Data, http://www.bis.org/bcbs/qisoprisknote.pdf (accessed 11 January 2010).

Basel Committee on Banking Supervision (2006) Basel II: International Convergence of Capital Measurement and Capital Standards: A Revised Framework – Comprehensive Version, http://www.bis.org/publ/bcbs128.htm (accessed 6 March 2010).

Ben Gal, I. (2007) Bayesian Networks, in *Encyclopaedia of Statistics in Quality and Reliability*, F. Ruggeri, R.S. Kenett and F. Faltin (editors), John Wiley & Sons, Ltd, Chichester.

bnlearn Package (2008) http://cran.dsmirror.nl/web/packages/bnlearn/index.html (accessed 21 May 2010).

Charniak, E. (1991) Bayesian Networks without Tears, *AI Magazine*, 12, pp. 50–63.

Cornalba, C., Kenett, R. and Giudici, P. (2007) Sensitivity Analysis of Bayesian Networks with Stochastic Emulators, *ENBIS-DEINDE Proceedings*, Turin, Italy.

Cowell, R.G., Dawid, A.P., Lauritzen S.L. and Spiegelhalter D.J. (1999) *Probabilistic Networks and Expert Systems*, Springer, New York.

GeNIe (2006) Decision Systems Laboratory, University of Pittsburgh, http://genie.sis.pitt.edu (accessed 25 January 2010).

Harel, A., Kenett, R. and Ruggeri, F. (2008) Modeling Web Usability Diagnostics on the Basis of Usage Statistics, in *Statistical Methods in eCommerce Research*, W. Jank and G. Shmueli (editors), John Wiley & Sons, Inc., Hoboken, NJ.

Heckerman, D. (1995) *A tutorial on learning with Bayesian networks*, Microsoft Research Technical Report MSR-TR-95-06, http://research.microsoft.com (accessed 11 January 2010).

Hugin Decision Engine (2007) *Hugin Expert*, http://www.hugin.com (accessed 6 March 2010).

Jensen, F.V. (2001) *Bayesian Networks and Decision Graphs*, Springer, New York.

Kenett, R.S. (2007) Cause and Effect Diagrams, in *Encyclopaedia of Statistics in Quality and Reliability*, F. Ruggeri, R.S. Kenett and F. Faltin (editors), John Wiley & Sons, Ltd, Chichester.

Kenett, R.S. and Salini, S. (2009) New Frontiers: Bayesian Networks Give Insight into Survey-Data Analysis, *Quality Progress*, August, pp. 31–36.

Murphy, K.P. (2001) *A Brief Introduction to Graphical Models and Bayesian Networks*, http://www.cs.ubc.ca (accessed 21 May 2010).

MUSING (2006) IST-FP6 27097, http://www.musing.eu (accessed 21 May 2010).

Pearl, J. (1995) Causal Diagrams for Empirical Research, *Biometrika*, 82, pp. 669–710.

Pearl, J. (2000), *Causality: Models, Reasoning, and Inference*, Cambridge University Press, Cambridge.

11

Combining operational risks in financial risk assessment scores

Michael Munsch, Silvia Rohe and
Monika Jungemann-Dorner

11.1 Interrelations between financial risk management and operational risk management

Financial risk management (FRM) and operational risk management (OpR) are strongly related in terms of data, methods and results. Information that is necessary to handle financial risks and credit ratings is also relevant for OpR, that is, it addresses information and monitoring of portfolio structures. We show a use case of the rating industry and an example of fraud protection. The same methods and data are used to estimate potential losses in this area.

Efficient FRM uncovers all risks in a comprehensive way and provides appropriate measures to minimise them. Risks must be quantified by their probability of occurrence and loss given default. This requires the construction of internal early warning systems.

Ratings of counterparties are an integral part of credit rating and FRM processes. For creditors, especially providers of financial services, there is always the possibility of having a firm with a risk of default in the customer portfolio. For this reason, a standardised and objective judgement of the analysed firm's ability to fulfil financial obligations on time is necessary. It makes contingency

Operational Risk Management: A Practical Approach to Intelligent Data Analysis   Edited by Ron S. Kenett
and Yossi Raanan   © 2011 John Wiley & Sons, Ltd

risks comparable and provides a selective management of the credit portfolio risks. A distinction is made between identification, measurement, management, limitation and reporting of the relevant risks.

Delayed payments and defaults of debtors are risks that are especially important for FRM. The quality of the used address data has, in case of failure of an amount of debit, a high relevance for possible collection and liquidation.

Ongoing and full information on existing and new customers is required for proactive credit risk management. This is the basis for standardised rating models and decision rules. A computer-aided data system for automated information procurement and implementation of rating models can have different special advantages. The credit management process is represented primarily as an information management process. Because of this, the credit grantor has to find possibilities to evaluate the risks of a credit in a fast, reliable and computer-aided way.

Rating and scoring systems are common tools for decision making in credit agreements. Rating is an evaluation method to represent the business situation of a firm. A better rating implies a higher rating class and therefore a firm's ability to make a payment on time. Ratings are the result of a rating process summarising the complex coherences of a credit evaluation. Ratings are classified as either internal or external. External ratings are provided by rating agencies. The rating reflects the statistical probability that the analysed firm will experience a breach of a payment obligation.

11.2   Financial rating systems and scoring systems

The term 'rating system' includes all methods, processes, controls, data collection and data processing systems used to determine credit risks, to assign internal ratings and to quantify losses. To use a rating system as specified by Basel II, for calculating capital requirements, demands that the rating scale consists of at least seven rating classes for the 'good debtors' in corporate loans. A two-level construction principle is used for this purpose. As a first step, a credit score value is assigned to debtors, which generates credit ranking. Following that, losses are allocated to a one-year period for different scoring intervals. The result is the so-called prediction profile. The second step includes a clustering of the firms into homogeneous classes in order to achieve the requested number of rating classes with adequate properties. So, to develop a rating system, a scoring system has to be developed first. Usually there is a need for:

- a useful distribution of firms across the risk classes with no excessive concentration in single classes (regarding gradation between debtor ratings and ratings of credit volume);

- exactly termed default definition, rating processes and criteria leading to a plausible and meaningful risk segmentation;

- a representation of the ratio between debtor risk classes as graded risk contents per risk class, meaning that the measured risk from one risk class to another rises equally with the decrease of the credit quality.

The rating strategy has to explain the risk of every class in the form of a representation of the probability of default allocated to all debtors of one risk class. A scoring system is a mathematical statistics process used to evaluate and respond to requests for credit in a standard or retail business. It also helps predict future payment behaviour. The goal is to optimise and rationalise credit judgements via the individual consolidation of internal and external data about debtors. A scorecard allocates individual items (i.e. mode of payment) to point values which add up to the score value. It is important that the individual items provide a good estimate of the probability of default.

Basically, scoring systems can be used throughout the process chain integrating customer and risk management. Audience targeting, marketing and mailing campaigns can start with marketing scores. Credit processing and FRM use application and behaviour scorings to evaluate credit applications and payment behaviour.

The idea of credit management, as a part of FRM, is to uncover data structures which are typical of specific behaviour of customers, and derive guidance out of this context. An important requirement is the so-called Stability–Time Hypothesis: a customer probably will behave similarly to comparable customer structures from the past. The development of the scorecard is based on mathematical statistics methods identifying the most selective items. If there are no historic cases available, generic scorecards are developed on the basis of expert knowledge and representative data.

In doing so, there are two potential sources of error: The α-error rates of future insolvency correspond to firms with an erroneous score of 'solvent' which eventually results in losses. The β-error rates represent future solvent firms classified as insolvent. However, error minimisation is difficult. A minimisation of the α-error to avoid insolvency losses usually causes an increase of the β-error and a higher rejection rate. Conversely, a very low risk preference minimises the β-error and the number of rejections while causing a higher number of insolvency cases (α-error). This can be described by the shift of the cut-off point in Figure 11.1. A good scoring system features a minimal α/β-error distribution for different settings and risk preferences.

According to the rules of Basel II with minimum requirements, the FRM scoring systems need to be permanently capable of ensuring high-quality predictions. Predicted loss rates should be reviewed with the help of extensive back testing. An important part of this process is the evaluation of stability based on statistical significance tests and the prediction of the quality of scoring with the Gini coefficient.

The Gini coefficient is a relative concentration measure with a value ranging from 0 (uniform distribution) to 1 (total concentration). This way, the Gini coefficient allows statements about the advantage of the scoring system in predicting a probability of default in comparison with an incidental finding, in percentages.

As a general rule, more losses from bad risk classes provide a better selection of the scoring and vice versa. Table 11.1 shows the distribution of losses in risk classes. In the calculation in the tables, above 55% of all losses at the beginning of the period under consideration were rated with the worst risk class. There are

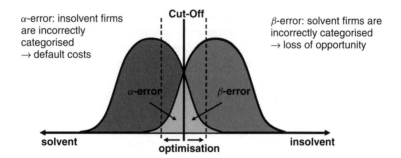

Figure 11.1   The α/β error distribution. This distribution displays which percentages of the firms are categorised in one of the two classes (good/bad) by mistake. Both errors are directly connected to each other. For example, the optimisation of the α-error causes a degradation of the β-error and vice versa.

Table 11.1   Distribution of losses in risk classes.

Risk class	Equipartition of the score population (%)	Distribution of all failures after one year (%)
1 (good)	10	0
2	10	0
3	10	1
4	10	1
5	10	2
6	10	2
7	10	5
8	10	8
9	10	26
10 (bad)	10	55
Total	**100**	**100**

also no observed losses in risk classes 1 and 2. Therefore it is a good working scoring. Without a scoring at the beginning of the year to evaluate requests and a full acceptance of every case, the losses would be distributed over all 10 classes.

For a scoring system in the retail business, a Gini coefficient starting at 45% is 'good' while 55% is 'very good'. Typical banking internal rating systems, processing annual closures, payment and account data as well as other quality items, show Gini coefficients up to 70%.

11.3   Data management for rating and scoring

The quality of a scoring system is mainly determined by the extent of the available raw data. A huge amount of data does not necessarily lead to a selective scorecard. The ability to filter the best data and make it usable is crucial.

Three important factors, with different impact levels, define the extent of the data, that is the number of items in the system. Early identification of relevant items is essential to control the effort of data collection. In complex systems, it is useful to analyse individual processes. Not every technically designated item can be used. Redundant items should be recognised quickly and excluded from further analysis.

The second value is the size of the data set. A balanced ratio between effort and selectivity is typically achieved with a number of about 10 000 usable data sets, the minimum quantity being 1000 data points. Discordant values require an exact investigation regarding the distribution of the characteristic items. Having more than 50 000 data sets does not add significant value to the quality of the scorecard.

The temporal horizon defines the third value. The purpose of a scorecard is to predict future developments on the basis of historic data. Therefore the data should not be too old. Data sets with an age of more than 10 years can hardly represent current occurrences and are not useful for scorecard developments. The temporal component should not be too short either, because individual years can show short-dated variations with a disproportional negative weight. Data from the last three to five years represents a practical middle course.

Another critical factor for the quality and practicability of the scoring system is the choice of the analysis platform. The decision is based on technical possibilities and the amount of data. A database server is counted as optimum. A matching concept is necessary in the case of raw data separated across multiple tables. The data is joined in an all-in file with the help of predefined identification keys. A system reorganisation during the enquiry period is a possible problem for such a join. Data migrations are especially not properly described in closed cases under given circumstances. More complex IT systems result in a costlier setup for an analysis database.

The available data is cleaned up in a following step. Incomplete or incorrect data sets can expand to disturb the whole upcoming development in a negative way. Duplicates should be eliminated, too. An extensive descriptive analysis should follow the join of all available data. Analysis of the filling degree as well as of the frequency distributions of the items gives results about the applicability of the items.

During the scorecard development some selective items have to be generated from the raw data. Some of these items – usually items like the amount of order or mode of payment – can be directly used in the scorecard. Others need a simple transformation, like date of birth into age of customer. The raw data offers additional key figures like the duration of a business connection or the average ordered sum. Such newly generated items can improve the selectivity of a scorecard. There are two factors to keep in mind: the key figures should make economic sense and therefore be relevant for evaluation of the probability of default to ensure the acceptance of the system. Additionally, key figures created in a development environment are often only usable in practice with great effort. Even the item 'amount of transacted businesses' can represent such a case, because the whole database has to be searched for data sets of the same customer.

A similar problem results in the usage of external data. It has to be ensured that this data is available anytime without delay.

After completion of the scorecard, the validation and a possible recalibration are highly important for controlling the prediction quality and to react to changes in the data sets. The scoring result is given a reference indicating the quality of the prediction. Equally, it is advisable to check the stability of the individual items. The validation database offers after a short period of time the basis for a recalibration of the scorecard.

11.4   Use case: business retail ratings for assessment of probabilities of default

Extensive credit management reduces depreciation on credits and leads, in the medium term, to an improvement of the equity situation of the creditor. The firm itself creates a positive awareness for its own refinancing with a good, documented risk management system. Essential parts of the credit management process in praxis are scoring systems that are based on external business information and standardised business retail ratings based on business information. They form another group of ratings next to external ratings and internal bank ratings and can be used as a basis for scorecard development.

Business retail ratings are based on credit information of independent credit agencies. In contrast to internal and external ratings, their advantage is the fact that basically every firm has such a standardised rating. In addition, such information is published, which is not standard in agency ratings. The external business information contains a variety of single pieces of information about the structure and financial situation, revenue and normative credit evaluation which allow an extensive assessment of the firm.

The concretisation of the two levelled construction principles will be described on the basis of the business retail rating of Creditreform. It is based on the solvency index, which is a central component, and scoring of the Creditreform credit report.

The solvency index allows rapid evaluation of a firm's creditworthiness. It summarises all relevant information to a value of the firm as a three-digit number. It is IT supported, calculated and composed from a variety of information describing the solvency of the firm. A statistical method, the linear discriminant function, is used to apply the single risk factors in a score value. Thus 15 hard (quantitative) and soft (qualitative) risk factors are weighted by their relevance in percentage terms. In detail these are: mode of payment behaviour, credit judgement, order situation, business development, employee structure, revenue, productivity as revenue per employee, equity structure, payment behaviour of the customers, payment behaviour of the firm, asset turnover, legal form, age structure, shareholder structure and industry situation.

The mode of payment behaviour is adopted by 25 to 29% and the individual Creditreform credit evaluation is adopted by approximately 25% in the

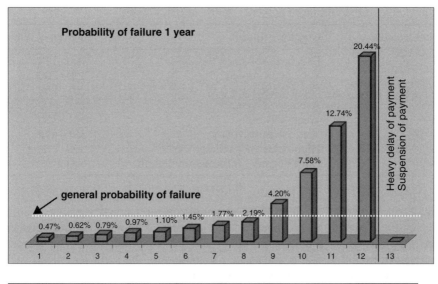

Figure 11.2   Failure probability for 12 rating classes (source: Creditreform).

calculation of the solvency index, depending on the firm's legal form. The ownership structure and environment of the firm and the previous business behaviour are considered in the credit evaluation. Revenue and financial data slip in with approx. 25%, structure data of the firm with 10 to 15%. The final 10 to 15% of the weights are part of the industrial sector and size-specific items. Then the weighted factors are summarised to a total value. A higher score value of the solvency index indicates a higher loss risk for the customer. The solvency spectrum ranges from 100 for very good solvency to 600 according to German school grades. For the case score value of 600 there are hard negative items present. Figure 11.2 below emphasises the importance of the single solvency classes.

Table 11.2 displays the values of the solvency index by risk class. The solvency index is not applied directly as a measure of creditworthiness in business retail ratings. Instead the idea is that the index is used to order the firms and also to define ranges of scores as rating classes. In practice, firms are ordered by the index at the beginning of a year and are compared with failed firms to determine the probability of default (PD). The probability of default for every rating class is defined by grouping the debtors into at least seven disjoint rating classes (as specified in Basel II) that cover the whole rating spectrum.

This inter-temporal pooling is in line with the Basel II requirements on PD estimates. Such estimates have to represent a long-term average value of the actual debtor's default rate in one calendar year and one rating class. The length of the observation period has to be at least five years. The fact that the probabilities

Table 11.2   Distribution of solvency index in risk classes.

Risk classes	Creditworthiness	Solvency index
1	Excellent solvency	100–149
2	Very good solvency	150–200
3+	Good solvency	201–250
3	Average solvency	251–300
3–	Weak solvency	301–350
4	Very weak solvency	351–499
5	High delay of payment	500
6	Hard negative information/default	600

of default generated by this rating are monotonically non-decreasing, with a decreasing creditworthiness, meets the requirements of Basel II. The link between the Basel II standard default definition and the Creditreform solvency index ranging from 500 to 600 is emphasised by observing the variables behind such an index value: (1) significant delay of payment up to 90 days, (2) several overdue notices, (3) temporary refund, (4) no fulfilment of dated commitments, (5) calling in a debt collection service and (6) insolvency procedures.

The average default rate in Germany is about 2.2%. This value is higher than the number of insolvencies stated in public statistics because of events like personal default of the owner of a small firm, besides insolvency cases. This demonstrates the quality of the solvency index. In the best solvency class, 0.47% of the firms fail. Firms in the worst class already show obvious signs of imminent creditworthiness degradation. The focus is especially on debt collection items.

With the help of the solvency index, the credit analyst is able to measure the prediction of default risks. Here the Gini coefficient allows evaluation of the prediction quality, too. Figure 11.2 shows the probability of default for 12 rating classes derived from the solvency index.

The application of business retail ratings supports the development of individual scoring systems in a positive way. Besides the solvency index, other indicators like payment behaviour are often used. Therefore the solvency index, the business retail ratings and the information data are central parameters for scoring systems.

An additional advantage for the credit decision-making process is fast and stable software-to-software communication between the dialogue system and the central system of the agency. In practice, this speeds up the decision process. The data is received in a structured and encoded form supporting automated processing by the scoring system implemented in the application processing (see Figure 11.3). Additionally, there is the possibility to build a customer database with all agency information so as to ease the information processing.

In the best case, the implementation of the scoring system or the business retail ratings into the decision-making process leads to a field service's ability to

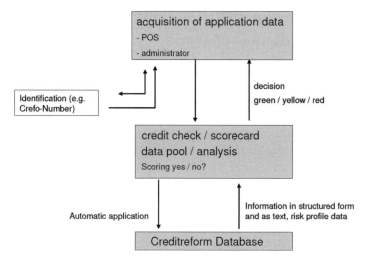

Figure 11.3   Automated scorecard application.

identify potential business partners via the database of the agency. In this way, a high degree of information and risk evaluation is guaranteed – even before signing a debt contract.

Depending on development techniques, data sets and statistical evaluation (monitoring, validation and back testing), scorecards vary in terms of quality. An external certification evaluating scorecards through benchmarks can remedy this situation. Therefore the following steps are necessary:

- provision of the customer portfolio data and scoring results for a defined time period;
- data preparation and data cleaning as well as an enrichment with external data;
- coordination of the 'good/bad' definition and separation of the portfolio on this basis;
- determination of the individual scoring system prediction quality with the help of selective measures (e.g. Gini analysis);
- determination of the α/β-error curve in addition to other analyses.

The effective use of scorecards requires a high acceptance level by employees. External approval of the system's prediction quality can help achieve this. Scorecards strengthen external communication with business partners, investors, banks, trading partners and customers and provide increased confidence in the firm. Overall, implementation of management's requirements in and internal FRM can be greatly helped by using the expertise of a well-known external agency.

11.5   Use case: quantitative financial ratings and prediction of fraud

In general, financial statement data has a special relevance for solvency judgements. Financial statement data has an important role in credit and supplier management of companies. It is also used to evaluate large businesses. There is criticism of solvency evaluations and ratings that are only based on financial reports, because they are oriented towards the past and affected by accounting policies. Assuming a minimal stable business development over time, there is the possibility to draw conclusions from key values on the upcoming development of a business partner. Numerous statistical studies emphasise the very good selectivity of rating systems based on balance sheets.

A quantitative financial rating is a rating system offering an evaluation of the probability of default based on a financial report. A statistical model, similar to a scoring system, underlies the evaluation. Figure 11.4 shows the basic structure of the model. The Creditreform financial rating uses a definition of failure that conforms to Basel II and has a very high predictive power with a Gini coefficient of almost 70%. Every rating result is allocated to a period-ordered failure probability. The failure probability rises from 0.07% in the class CR1 up to more than 15% in the class CR19. This is the one-year failure probability. The failure probability can be calculated up to a period of five years (see Figure 11.5).

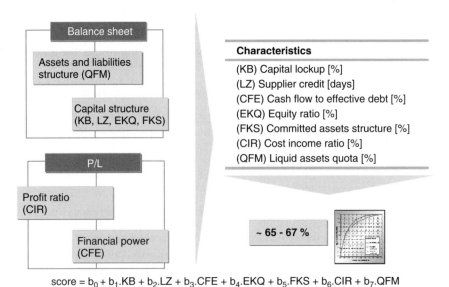

$$\text{score} = b_0 + b_1.KB + b_2.LZ + b_3.CFE + b_4.EKQ + b_5.FKS + b_6.CIR + b_7.QFM$$

Figure 11.4   Balance sheet analysis.

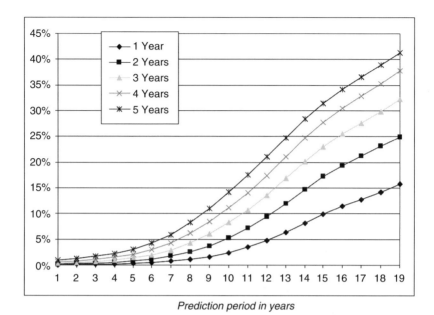

Figure 11.5   PD curves for different periods of prediction.

It is obvious that, in the long run, the probability of default may increase up to 40% (five-year prediction in rating class CR19)

Financial ratings are used primarily as reliable solvency indicators. An interesting challenge is to detect negative developments in a company with the help of financial ratings, despite fraud and balance sheet manipulations. Is it possible to recognise a general business crisis, from an external point of view, while there is internal balance sheet manipulation? In some cases it is possible, as shown below.

The link between fraud and business crisis reported in the literature invokes the hypothesis that the nature and impact of a business crisis possibly motivates fraud. By referring to an example, we show that quantitative financial ratings can detect a business crisis, even in fraud situations.

The Philip Holzman AG company went into default in 2002, after several years of balance sheet manipulation. Three years before the default, the rating was on CR15 with a one-year PD of 9.54% (see Figure 11.6).

This is only an example, but balance sheet manipulations were reported in several other cases of fraud. The detailed link between operational risks and financial ratings is still to be investigated by future theoretical and empirical research.

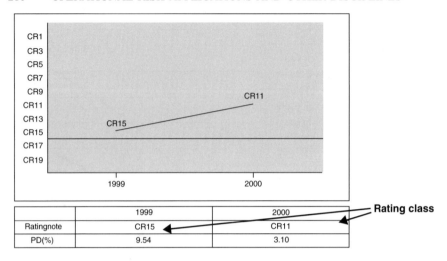

Figure 11.6   Development of the balance quality.

11.6   Use case: money laundering and identification of the beneficial owner

In the determination and control of risk, operational risks play a special role. Operational risks range over diverse divisions of financial service providers and are not always recognisable at first sight.

In particular, the frequency of fraudulent acts and transactions with criminal backgrounds, by sources external to the financial service industry, has risen in recent years. Criminal and terrorist organisations use different ways to legalise their illegally acquired 'dirty money' (e.g. from drugs, human trafficking, the arms trade and illegal gambling) or terror funding assets.

Therefore, prevention of such crimes gains in importance, with the 'know your customer' (KYC) principle being one of the most important instruments to negate money laundering and terror funding activities. The goal is to create transparent business relationships and financial transactions on a risk-oriented basis. Economic transactions by dummy or bogus companies should be identified and prevented.

This principle is extended by the new Money Laundering Act in Germany (*Gesetz zur Ergänzung der Bekämpfung der Geldwäsche- und der Terrorismusfinanzierung (Geldwäschebekämpfungsergänzungsgesetz – GwBekErgG)*), which became effective on 21 August 2008. This extension of existing statutory law has tightened the legal requirements on suppression of money laundering.

One of the major changes to the existing legislation is the breadth of financial instruments caught by the regulation and the enactment of the changes needed for the suppression of money laundering and terror funding.

One particular example of this is the extension of the economic beneficiary term (Paragraph 1 (6) Act of Detection of Proceeds from Serious Crimes Money Laundering Act – GwG). Paragraph 3 (1) (3) GwG demands that committed persons clarify if the contractual partner acts for a beneficial owner. This new definition is broader than the existing term in Paragraph 8 of the previous GwG.

For the purposes of this Act, 'beneficial owner' means the natural person(s) who ultimately owns or controls the contracting party, or the natural person on whose behalf a transaction or activity is being conducted and is applied in particular circumstances:

- In the case of corporate entities, which are not listed in a regulated market, the natural person(s) who directly or indirectly holds more than 25% of the capital shares or controls more than 25% of the voting rights.

- In the case of legal entities, such as foundations, and legal arrangements, such as trusts, whoever administers and distributes funds or arranges for third parties to administer and distribute funds:

    (a) The natural person(s) who exercises control over 25% or more of the property of a legal arrangement or entity.

    (b) The natural person(s) who is the beneficiary of 25% or more of the property of a legal arrangement or entity; where the individuals that benefit from the legal arrangement or entity have yet to be determined, the class of persons in whose main interest the legal arrangement or entity is set up or operates.

Listed companies and credit institutions are exempt from the term of economic beneficiary.

The following example, in Figure 11.7, illustrates the determination of the beneficial owner.

To identify the beneficial owner, authorised parties will be able to determine the share-ownership ratios by reviewing the information documents.

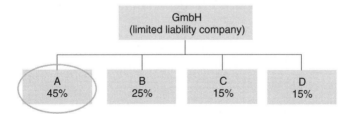

Figure 11.7   Identification of the beneficial owner. The beneficial owner is share-holder A. Shareholders B, C and D hold exactly 25% or less of the shares and therefore do not meet the requirements of a beneficial owner.

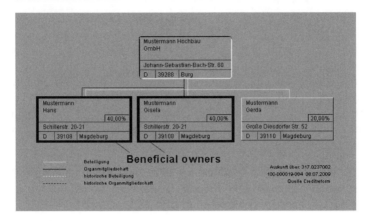

Figure 11.8   Integration of information (source: Creditreform).

Creditreform offers additional tools to identify the beneficial owner. The determination can be done manually with Creditreform's **credit reports** and integration information (Figure 11.8).

The application of a system of rules is obligatory for those persons using these products.

Financial service providers, like leasing companies, are used to working with automated processes. The determination of the economic beneficiary can be done with the help of the system on the basis of the 'CrefoSystem' customer platform and therefore facilitate workflows in business processes (see Figure 11.9). Every transaction, every system of rules and every interaction with the user are documented and chronologically recorded.

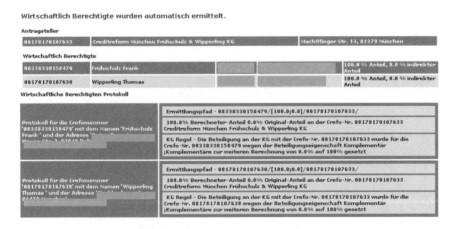

Figure 11.9   A sample automated initial report from the CrefoSystem system (source: Creditreform).

11.7   Summary

Risk indicators are measureable values indicating a statistical link between specific risks. They reveal an increased probability that a risk is going to happen and at the same time allow for its prevention. An early warning system monitoring these indicators can show negative developments and trends, permitting counteractions to be taken in a timely manner.

Operational risks also need a definition of adequate and risk reflecting indicators. Overall, the aim is to identify a connection between typical financing processes, operational risks and useful risk indicators. Risk indicators should meet the criteria of transparency, objectivity, profitability and relevance. They have to be measureable in an early, regular and efficient way. Additional characteristics are uniqueness, completeness and the possibility of adding thresholds for counteractions.

Despite these high requirements for developing appropriate indicators, there are several meaningful expressive key values matching operational risks. The control of the consulting processes in the car leasing business is one example. The frequency of fraud by external persons in the leasing business has risen in the last few years, especially fraud by dummy corporations and shell companies. The challenge for the control of the leasing and handling processes is not to disturb the automated evaluation processes in routine business while optimizing point of sale procedures with a focus on fraud detection.

The risk in concluding a leasing deal is based on several factors like the creditworthiness of the debtor and the capability of the company's own internal evaluation processes. Appropriate indicators can help to identify systematic developments at an early stage. The use of Internet-based verification tools from trusted sources for the evaluation of customer data like bank account or address verification, and scoring modules like small ticket leasing, should be emphasised. During the request evaluation, the credit reports can deliver data-driven early warning indicators. The aim is to concentrate 85 to 90% of all fraudulent business processes in 5 to 10% of the transactions by using intelligent filtering processes. Exceptions that cannot be verified by automated processes have to be checked by special examination.

The determination of a first selection and identification of risk factors for operational risks in self-assessment should be done by business and risk managers. Participating employees are asked about the key values they would monitor regularly to make a statement about the described risk scenarios.

Here are some examples for useful risk indicators:

- duration and number of system breakdowns;

- capacity and occupancy rate of storage media, processors and network;

- number of vacant positions;

- disease and fluctuation rate;

- number of customer complaints;

- reversal rate;

- key values of customer, supplier and service provider ratings;

- budget for further training.

In many cases, only a combination of several indicators and key values provides meaningful indices for the development of the current risk situation. Every firm is requested to extend and refine the early warning system on the basis of collected empirical values. One part of this process is the implementation of threshold values as some kind of tolerance limit showing entry into a critical sector and triggering an escalation mechanism of identified indicators and key values. Risks with an immediate current need for action have to be part of the reporting, economically useful and evaluated regularly.

The chosen indicators have to be transparent and integrated with internal and external measures of the firm during implementation of the described early warning system. In this case, the reporting should not consist of an avalanche of numbers or 'number graveyards' – the motto is 'less is more'. Other requirements for a successful implementation are data availability and the technological environment.

Firms should establish an early warning system within the scope of their FRM processes despite existing implementation problems in practical use. Besides the legal and supervisory requirements, there are clear economic advantages, like the prevention of fraud, to be gained. For more on these topics, see Basel Committee on Banking Supervision (2006), Dowd (1998), Englemann and Rauhmeier (2006), MUSING (2006) and Tapiero (2004).

References

Basel Committee on Banking Supervision (2006) Basel II: International Convergence of Capital Measurement and Capital Standards: A Revised Framework – Comprehensive Version, http://www.bis.org/publ/bcbs128.htm (accessed 21 May 2010).

Dowd, K. (1998) *Beyond Value at Risk: The New Science of Risk Management*, John Wiley & Sons, Ltd, Chichester.

Engelmann, B. and Rauhmeier, R. (2006) *The Basel II Risk Parameters*, Springer, Berlin.

MUSING (2006) IST-FP6 27097, http://www.musing.eu (accessed 21 May 2010).

Tapiero, C. (2004) *Risk and Financial Management: Mathematical and Computational Methods*, John Wiley & Sons, Inc., Hoboken, NJ.

12

Intelligent regulatory compliance

Marcus Spies, Rolf Gubser and Markus Schacher

12.1 Introduction to standards and specifications for business governance

Today, enterprises have to cope with a large volume of regulations. The amount of regulation, the type of regulation (financial conduct, anti-terrorism, privacy, employment, health care, environment, food and drugs, minority rights, taxation, safety, etc.) and the number of regulators increase year by year. It is difficult to understand the full extent of such regulations, how it is relevant to the company and how to keep up to date with all of it. As regulations come from many sources, they may overlap in scope or even conflict in their requirements. Aggravated by the fact that regulations are often ambiguous, they have to be interpreted. The definitions of terms are often not sufficiently precise to formulate concrete compliance policies or they might even mean different things in different, related contexts.

The effect of this is that regulatory compliance is a major overhead for business, and enterprises need better standards, better tools and more coordinated approaches to cope with it. Regulatory compliance management is a key issue and it has different aspects:

1. Interpreting what regulations mean to a specific enterprise.

Operational Risk Management: A Practical Approach to Intelligent Data Analysis   Edited by Ron S. Kenett and Yossi Raanan   © 2011 John Wiley & Sons, Ltd

2. Assessing the impact of the regulations on the enterprise's policies and operations.

3. Deciding how to react, and establishing policies and guidance for compliance.

4. Demonstrating that compliance policies are being followed across the enterprise, that controls are implemented and that they are effective.

Regulatory compliance management is a core function of business governance – it is not just about risk management. Today, a good framework for business governance has to integrate different disciplines like strategy planning, business process management, policy management and internal control in a consistent and well-integrated management system. Such a management system should:

- enable the sharing of business knowledge (inside the enterprise: stakeholders, board, senior management, operational managers – but also outside: external auditors, partners, regulators);

- provide a holistic view of the enterprise;

- enhance collaboration between departments, roles and responsibilities;

- enable automation, as much as possible;

- be easy to keep up to date with internally or externally required changes.

To implement such a management system, an organization needs:

- a well-defined structure to manage complex information about the enterprise;

- adequate tools that work on this structure and provide a high degree of automation – down to operational IT systems.

The Object Management Group (OMG), one of the world largest standardization organizations for business software and software architecture, defines several specifications on how to structure and outline business models (OMG, 2010a). The business motivation model (BMM), introduced below, is such a specification. It can be seen as a framework for implementing a management system for business governance, and, as such, it is designed to cope with complex business knowledge in a volatile environment.

A model-based approach builds the foundation for intelligent regulatory compliance (IRC). By IRC, we mean semi-automatic or fully automated procedures that can check business operations of relevant complexity for compliance against a set of rules that express a regulatory standard. Thus, IRC supports the implementation of business governance in a context of increasing regulatory requirements.

The concept of IRC is also meant to be understood in the context of automation of business governance implementation. We describe below the relationship between IRC and business rules. For business rules, automation frameworks are becoming more and more relevant due to the bundling of business rules engines with business process management software. While existing solutions to compliance management mostly belong to the IT application domain of business performance management, IRC opens up a perspective to *regulatory compliance by design* – meaning that compliance checks are integrated with the definition and deployment of business processes.

IRC is of particular relevance to risk management. The approach presented here was developed in the context of the Basel II regulatory framework for capital adequacy (Basel Committee on Banking Supervision, 2003). Basel II distinguishes between financial, market and operational risks. For financial risk management, IRC can support auditing and review processes that qualify financial services implementing the internal rating-based advanced measurement approach (see Chapter 10). This is seen by many banks and financial institutions as a key advantage compared with externally defined and audited risk measures. In MUSING, a virtual network operator (VNO) and the MAGNA system of the Israel Securities Authority cooperated in the implementation and testing of the IRC research presented in this chapter (MAGNA, 2010; MUSING, 2006).

12.2 Specifications for implementing a framework for business governance

We base our approach on the framework for business governance as developed by OMG. OMG's recent specifications for business planning and business design support implementation of an integrated framework for business governance.

More specifically, we will discuss how the BMM (Business Motivation Model) and SBVR (Semantics of Business Vocabularies and business Rules) specifications by the OMG provide a suitable basis for representing regulation systems in a sufficiently formal way to enable IRC of business process models.

Regarding compliance management in general, OMG is currently working on different topics:

- *OMG-RFP: Management of Regulation and Compliance (MRC)*
  A BMM and SBVR based specification that aims at defining suitable process and reporting metadata for regulatory compliance using the eXtensible Business Reporting Language (XBRL).

- *The Governance, Risk Management and Compliance Global Rules Information Database*
  (GRC-GRID, 2010b, www.grcroundtable.org/grc-grid.htm)
  The Governance, Risk Management and Compliance Roundtable (GRC-RT) is developing a Global Rules Information Database (GRC-GRID or GRID) as an open database of rules, regulations, standards and government guidance documents that require IT action and a survey of the regulatory climate around the world. The goal of this project is to provide the de facto GRC reference guide for global IT and business managers. The initial development of the GRID is complete and governed by the GRC-RT.

The GRC-RT promotes its use, guides and implements its enhancement and collaborates with rules producing entities worldwide to oversee and automate data acquisition.

12.2.1   Business motivation model

BMM provides a scheme or structure for developing, communicating and managing business plans in an organized manner. More specifically, BMM does all of the following:

- It identifies factors that motivate the establishment of business plans.
- It identifies and defines the elements of business plans.
- It indicates how all these factors and elements interrelate.

It can be used to:

- develop business plans;
- manage business plans in a volatile environment including the definition of suitable indicators or business scorecards;
- document business strategies and operational policies;
- demonstrate compliance (SOX, Basel II, etc.) of real business processes with regulatory requirements provided by business policies and directives.

Among these elements are those that provide governance for and guidance to the business – business policies and business rules.

BMM is a general model for managing change (see Figure 12.1). A BMM for an enterprise supports a control system for business operations, with four aspects:

1. Monitoring influencers that affect the business.

2. Assessing the impact on the business of changes caused by influencers. Influencers may be internal to the business, or external.

3. Deciding how to react – what means to adopt. Means include:
   (a) Courses of action: strategies and tactics for using resources and assets. Courses of action are realized in the operational business by organization roles and responsibilities, business processes and business rules.
   (b) Directives: policies and rules that govern the courses of action.

4. Deciding how to measure the effectiveness of the courses of action – defining desired results. Actual values will be obtained from the operational business for comparison.

Regulation is one category of external influencer that can be managed with an enterprise BMM. It is important to note that BMM is not in any sense a

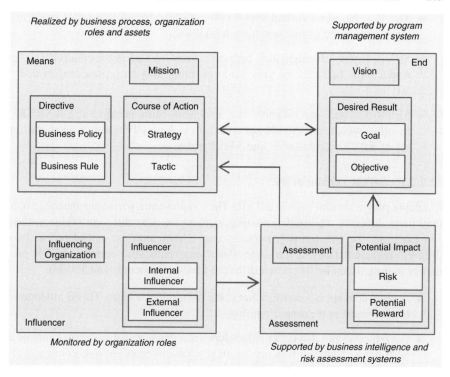

Figure 12.1   Overview of BMM.

methodology. Indeed, it is entirely neutral with respect to methodology or a particular management approach. One way to think of BMM is as a blueprint designed to support a range of methodological approaches. Implementation of BMM would result in the elements of business plans being stored and related to other information about the enterprise, no matter what methodology was used for discovering and defining them.

12.2.2   Semantics of business vocabulary and business rules

Another important specification that is essential to enable IRC is SBVR. This is a specification developed by the OMG and published in its first version in January 2008 (OMG, 2008a). It specifies how vocabularies of business concepts shall be expressed and how such vocabularies shall be used to formulate business rules. It is based on the following fundamental principles:

- SBVR is a domain-independent specification and primarily addresses the business community (i.e. it is not an IT specification).

- SBVR provides explicit support for multiple communities (i.e. different groups of people that share some common knowledge or groups of people that share a common language (semantic communities)).

- SBVR is based on formal logic, more specifically on first-order predicate logic with some extensions from modal logic.

- SBVR strictly distinguishes between meaning and representation (i.e. it respects the fact that the very same meaning may be represented in many different ways).

- SBVR is a 'self-specification' (i.e. its specification is expressed in SBVR). In other words, the SBVR specification is basically nothing more than a set of SBVR vocabularies and SBVR rules.

12.2.2.1 SBVR vocabularies

Concepts play a central role in SBVR. They represent a particular meaning that may have different representations (e.g. names) used by different communities. For example, the two terms 'customer' and 'client' may be considered as two different representations of the same underlying (nameless) concept. SBVR primarily distinguishes between two different kinds of concepts (meanings):

- A noun concept is a concept that is the meaning of a noun. This is analogous to a concept in a domain ontology.

- A verb concept (also called a fact type) is a concept that is the meaning of a verb phrase that involves one or more noun concepts. A verb concept with two roles corresponds to an object property or an object-valued relationship between concepts in a domain ontology.

Each concept in an SBVR vocabulary is described by a set of properties such as its representation (name) commonly used by a particular community and its precise definition, optionally complemented by potential synonyms, source, examples and a set of necessities (see the following section). The following fragment of an SBVR vocabulary shows an entry for a noun concept and one for a verb concept:

customer	
Definition:	A person that pays for goods or services
Synonym:	client
Source:	WordNet (adapted)
Necessity:	The name and address of a customer are known.
customer pays good	
Definition:	A customer that receives a good from us and gives us an amount of money in exchange for it.
Synonym:	good *is* paid *by* customer
Necessity:	A good *is paid by* exactly one customer.

Figure 12.2   SBVR concepts in graphical notation.

Alternatively, concepts within an SBVR vocabulary may also be represented in a graphical notation that is based on OMG's UML notation (Figure 12.2).

Finally, SBVR vocabularies may not only define their own concepts, but also adopt concepts from other common vocabularies.

12.2.2.2   SBVR rules

In SBVR, concepts defined in vocabularies are used to express rules, or, more specifically, business rules. In SBVR, a business rule is a 'proposition that is a claim of obligation or of necessity and that is under business jurisdiction'. This definition leads to the two fundamental types of business rules distinguished by SBVR:

1. A structural business rule is a 'claim of necessity under business jurisdiction'. A structural business rule is descriptive as it is used to describe the meaning of one or more noun concepts. For example, a structural business rule may state that 'a customer is a good customer, if he/she bought goods for more than €1000 [$1200]'. More specifically, structural business rules specify what is necessary, what is possible and what is impossible.

2. An operative business rule is a 'claim of obligation under business jurisdiction'. An operative business rule is prescriptive as it is used to prescribe how to conduct the business. For example, an operative business rule may state that 'a good customer must receive a 5% VIP discount on any purchase'. More specifically, operative business rules specify what is obligatory, what is permitted and what is prohibited.

An important difference between structural business rules and operative business rules is the fact that the latter may be subject to violation. Therefore, operative business rules must be enforced, that is their application must be monitored and any violation must be handled appropriately. In other words, we must ensure that the business behaves compliantly.

12.2.2.3   SBVR Structured English

In the previous section, business rules have been stated informally. SBVR also non-normatively specifies a simple language called 'SBVR Structured English' to express business rules more formally. Each sentence expressed in SBVR Structured English solely comprises the following word classes for which SBVR proposes different font styles:

- terms (<u>petrol, underlined</u>) and names (<u><u>petrol, double-underlined</u></u>) for noun concepts defined in a vocabulary;

- terms for verb concepts (*blue, italics*) defined in a vocabulary;

- predefined keywords (red, normal).

So, in SBVR Structured English the two business rules from the previous section would be represented as follows:

- It is necessary that a <u>customer</u> is a <u>good customer</u>, if and only if that <u>customer</u> *has bought* <u>goods</u> that *cost* more than <u>€1000</u>.

- It is obligatory that a <u>good customer</u> *receives* a <u>5%</u> <u>VIP discount</u> *on* a <u>purchase</u>.

These rule statements may be transformed to formal logic expressions that represent their meaning in an unambiguous form. In addition to SBVR Structured English, SBVR also shows how rule statements may alternatively be represented in other syntaxes based on natural language, like RuleSpeak® (BRS, 2010) and ORM (Halpin, 2001).

12.3 Operational risk from a BMM/SBVR perspective

BMM comprises a general concept of **risk**. Risk in BMM specializes the **potential impact** metaclass. The potential impact, in turn, motivates or provides impetus for **directives**. A directive, loosely speaking, implements a **regulation** in such a way that business processes, in the scope of the regulation, can be managed effectively through governing business policies or directly guided (on the operative level) by business rules. This sequence of associations is the basis of applying BMM in the context of regulatory compliance, as will be explained in more detail in the next section.

In the present section, we examine the concept of risk in some more detail, taking into account the highly influential Basel II framework for defining categories of risk and its foundation in statistical risk analysis (see Chapter 7). Risk is associated with a generic causal metamodel that comprises, at least, the following components:

1. Loss event types are defined on three levels of detail. As an example, business disruption is a general *category* comprising various system failures that can be further detailed as to the system type and exact failure description – this level of detail is often referred to as that of *causes*. To each loss event type in a specific enterprise or operational environment, probability distributions can be associated that capture expected frequencies over time or mean times to failure etc.

2. Business lines (or *domains*) are the 'place' where a specific risk is to be analysed.

3. Loss types describe the specific kind of loss incurred on a business line due to a loss event. Examples of loss types are write-downs, loss of recourse, regulatory and compliance-related losses (penalties etc.). A loss type determines the relevant indicators for loss measurement (*exposure* in economic risk analysis parlance). The measurements of these indicators are also said to quantify the *severity* of a loss event.

A key statistical approach in risk analysis is the convolution of causal and exposure probability distributions that yields an overall distribution of expected losses over time (or another domain used in the causal probability distribution). Using a specified percentile tail area under this convolved distribution, the so-called value at risk (VaR) at this percentage can be taken as a summary indicative number (see Chapters 2 and 7).

This risk metamodel has been implemented in the MUSING project using formal ontologies based on the Web Ontology Language (OWL). This is described in detail in Chapter 3 of the present book. The ontology has been used to formulate specific models for real-world prototypes implementing IT operational risk analytics, notably for TBSI, an Israel-based virtual network operator (see Chapters 5, 8 and 9), and for a data centre at the Monte dei Paschi di Siena Bank (see Chapter 10).

Basel II distinguishes three key areas of risk management – credit risk, market risk and operational risk. In the present chapter we focus on operational risk and assume an operational risk analysis in line with the causal metamodel to be in place at a given operational domain in an enterprise.

The causes of loss events are rooted in the activities of a business line, that is in a business process. A business process is, as we saw, guided by business rules and governed by business policies. If the analysis reveals a given activity as the cause of extremely severe losses, obviously a change in the business process is mandatory. To bring about such a change in a reasonably managed way, it is not sufficient to address the business process itself, for example by adding control activities. What is needed is an analysis of the governing policies and guiding rules of the process:

- The policy might need to specify additional directives – for example, in the area of distributed access control, operative losses in a network operation's centre might require changes in the policies for granting authorizations to users.

- The business rules might need additions/deletions – for example, in public transportation, the maintenance intervals for locomotives need to be shortened.

To sum up, in order to address these analyses and changes, the directive's element and its descendants in BMM are required. Figure 12.3 presents a first overview of the operational framing of a loss event due to operational risk in terms of regulatory constructs (rules, policies) derived from BMM. For additional references,

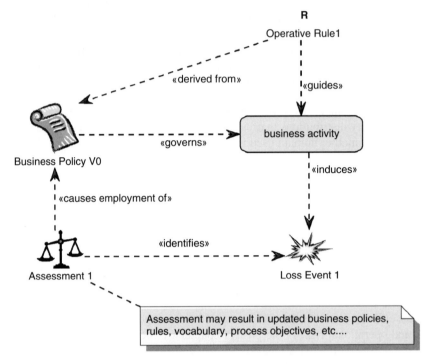

Figure 12.3   Loss events in context of BMM elements.

see EFIRM (2010) and MDE (2010). More details are provided in the remaining sections.

Since both business policies and business rules rely on suitable vocabularies, we also apply SBVR in this analysis and the subsequent processes of change management.

The relationship between BMM/SBVR and risk management can be even further extended to the strategic level. Many business operations are too complex to redesign them entirely and immediately in reaction to losses. Instead, what is required are mitigation strategies that allow soft transitions from given business practices to improved new ones. The strategic level of operational risk management comprises the setting of business goals and specific quantifiable objectives – again, this is exactly in line with BMM. On a strategic level, BMM introduces the (meta)model element CourseOfAction which may allow for different business processes to implement it or parts of it. Assuming that an enterprise faces substantial threats due to operational failures, a strategically planned course of action is needed that is then detailed in business processes. As an example, implementing a service quality strategy, a CourseOfAction could result from the tactical decision to generally revise all maintenance procedures in a set of owned power plants in line with a quality goal and quantified objectives. The maintenance procedures here are the specific business processes that need to be modified in accordance with revised directives. However, the directives are not immediately drawn from risk analysis in this example, but from a supervisory control process. From a practical point of view, this improvement cycle for operational risk management based on BMM/SBVR is often impossible to implement since today's business process modelling languages mostly do not explicitly refer to directives (policies, rules). While formal and executable business process description languages are specified and find their way into business process definitions (see the OMG BPMN – Business Process Modelling Notation – and WS-BPEL standards), the generation of process descriptions from a given set of business rules and governing policies has found less attention in research projects or standardization bodies, see the DecSerFlow approach by van der Aalst and Pesic (2006), and related model checking approaches to business process modelling (e.g. Hallé et al., 2007; Graml, Bracht and Spies, 2008).

12.4 Intelligent regulatory compliance based on BMM and SBVR

This section shows how standards and specifications like BMM and SBVR enable an approach for intelligent regulatory compliance (IRC). We illustrate the approach with a banking example using an advanced measurement approach (AMA) based on quantitative and qualitative data. It has five steps:

1. Assess influencers.

2. Identify risks (or potential rewards).

3. Develop risk strategies.

4. Implement risk strategies.

5. Build adaptive IT systems.

The SBVR diagram in Figure 12.4 shows the most important concepts of operational risk management and BMM used by the IRC approach. Among the concepts considered, we particularly focus on so-called near misses, multiple losses and opportunity losses. These are specific to loss events in the domain of operational risk and are defined as follows:

- Multiple losses: a common IT failure causes multiple consequences. In terms of business analytics, the problem is to infer the probable causing failure from the analysis of already recorded and apparently uncorrelated losses.

- Opportunity losses: quantify the missed business opportunities caused by IT failures, in terms of either forgone profits or greater expenses.

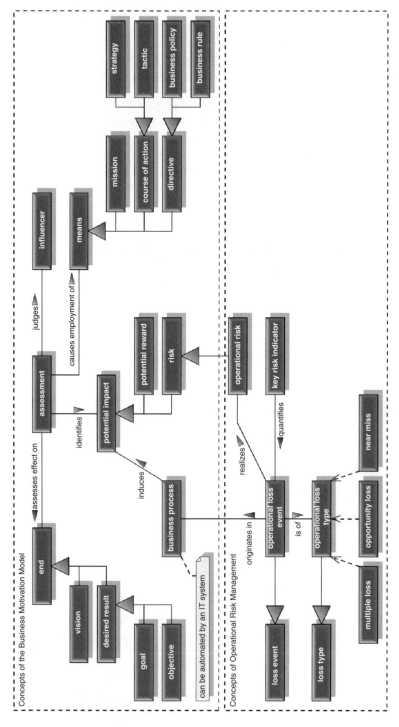

Figure 12.4   Interrelations of concepts of BMM and operational risk management.

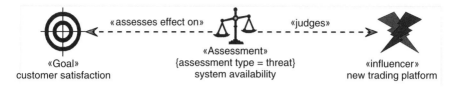

«Goal»
customer satisfaction

«Assessment»
{assessment type = threat}
system availability

«influencer»
new trading platform

Figure 12.5   Assessments in BMM.

- Near misses: track near misses from a KRI (Key Risk Indicator) perspective; near misses here are generically meant as operational risk events not producing a loss.

12.4.1   Assessing influencers

An important step in the process of risk identification is the assessment of influencers to the business – either external influencers, like new regulatory requirements, or internal ones, like resources or infrastructure. An influencer usually has an impact on the achievement of an end – in a positive or a negative sense. Thus it can be assessed by strategic analysis as a strength, a weakness, an opportunity or a threat. BMM makes those deliberations explicit and traceable for documentation or analysis purposes (see Figure 12.5).

For instance, if a financial services provider introduces a new trading platform, which is managed by an external IT provider, it has to assess the impact on the achievement of its objective to increase customer satisfaction. The resulting assessment identifies the threat – caused by the loss event of a system outage, which would have an effect on customer satisfaction, so that customers may be lost.

12.4.2   Identify risks and potential rewards

Risks or potential rewards are identified based on the assessments of influencers. Again, it is important to document assessments explicitly, as risks and potential rewards are often just different sides of the same coin (of an influencer). In our example, the assessment of the system availability of the bank's new trading platform has on the one hand identified the risk of missed trading orders (i.e. at first glance, an opportunity loss due to the loss event of a system outage, but it could also be qualified as a multiple loss event as soon as customers resign their contracts). On the other hand, a potential reward of fair trading provisions can be identified by assessing the effect on the objective of increased turnover (see Figure 12.6).

Besides the graphical representation of BMM, each element of such a model is stored in a database and has type-specific properties for its detailed description (e.g. fields for description, risk probability and impact for risk elements). Using

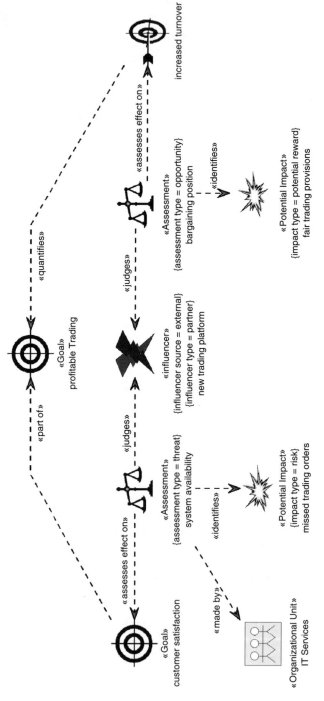

Figure 12.6 Risk identification with BMM.

BMM, the process of risk identification is well documented and pursuable – and for people who are not involved directly.

12.4.3   Develop risk strategies

Depending on risk impact and probability, adequate risk strategies have to be developed. In BMM terms, we have to employ some means (i.e. strategies, tactics or business policies) to define how to deal with an assessment. In our example, the assessment of system availability causes the employment of the trading service strategy, a risk mitigation strategy which is refined by three specific tactics.

On that level, business policies define the guidance for processes – carried out either by human workers or by IT systems (see Figure 12.7).

Finally we close the loop, from our risk strategy back to the initially assessed objective, to measure our success via the new objective of reduced system outages.

12.4.4   Implement risk strategy

Implementing a risk strategy means aligning the operational business in such a way that it realizes BMM's course of actions and directives. That means defining processes, assigning responsibilities, establishing control activities and audit trails. In this area BMM only provides the infrastructure to link the elements of the operational business – like processes, rules and organization – with strategies, tactics and policies. Typically, the operational business is modelled using BPMN for process descriptions and SBVR for business rules and their underling vocabulary.

Besides the operational business processes, specific processes, controls and rules for risk reporting (loss events, incidents and risk quantification) as well as for incident management have to be designed (see Figure 12.8).

12.4.5   Outlook: build adaptive IT systems

Using BMM, SBVR and BPMN as shown in our approach for IRC is more than implementing risk and compliance management in a systematic way. These specifications enable complex business knowledge to be captured in a form that is readable by humans and by machines. Thus it is also an excellent basis for building adaptive IT systems. Incidentally, that is a major goal of OMG's specifications.

With adaptive IT systems, changes in business processes and business rules can be implemented, to a large degree, without programming and independent of IT release cycles. That is useful not only in the context of risk management systems, but generally for all IT systems which have to act based on volatile and/or complex business knowledge.

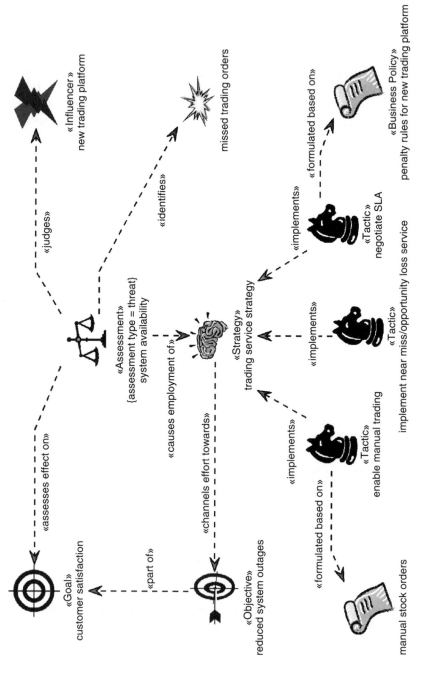

Figure 12.7   Defining risk strategies with BMM.

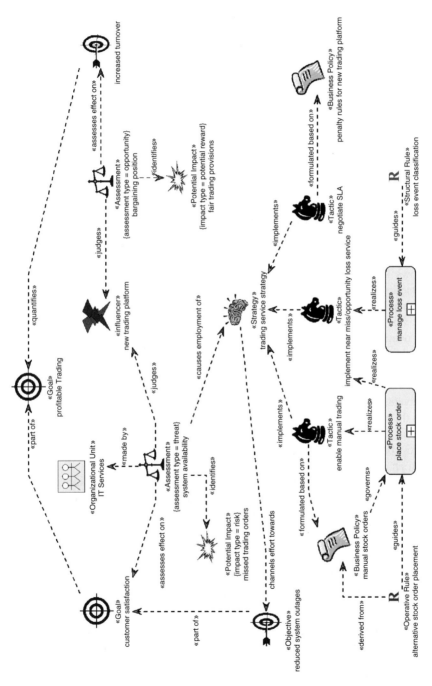

Figure 12.8 Linking operational business models with BMM.

12.5   Generalization: capturing essential concepts of operational risk in UML and BMM

This section presents a deeper analysis of the relationship between some generic UML concepts and operational risk concepts. The purpose is to establish a transformation framework that allows operational risk concepts to be expressed in terms of related UML modelling elements. This also addresses the integration of BMM and UML, and it must be noted here that this integration is still being discussed among the specification contributors. One benefit of a conceptual model of operational risk in terms of BMM integrated with UML is that model checking and constraint processing can be reused from UML tools and that UML-based code generation can be applied to sufficiently specified BMM models. This gives an additional perspective on integrated software engineering for IRC.

First, we present a UML metamodel capturing those elements of BMM that are mainly relevant to operational risk analysis, see Figure 12.9. This is condensed and adapted from a draft of the OMG BMM specification (2008b) task force. It shows key BMM concepts as metaclasses with suitable (meta-)associations. The model shows how the three key concepts of risk, regulation and business processes are connected by a framework consisting mainly of assessments and directives. This provides a comprehensive and yet succinct conceptual basis for more detailed domain models.

We now proceed to an integrative UML metamodel containing the essential BMM concepts plus specific operational risk entities and relationships, see Figure 12.10. The focus here is on defining the necessary concepts for describing loss events. Loss events belong to two different conceptual areas. On the one hand, they occur in business process instances and, as such, need to be related to UML activities in a plausible way. On the other hand, loss events are related to business performance management, and impact on one or more specific indicators within an assessment system for business objectives.

In setting up the integrative metamodel of loss events, we noticed that a proper UML 'framing' of the concepts and notions common to the IT operational risk community, including Basel II terminology, requires some adaptations and conceptual reanalyses, which we explain here in more detail:

- A loss event in operational risk usually can be analysed in terms of causes and consequences, thus it cannot be equivalent to an event in the UML sense, which is simply a communication behaviour.

- What is called in Basel II an 'activity' causing a loss event (like failure to report, theft, fraud, system failure) is not an activity in the UML sense ('Activity modeling emphasizes the sequence and conditions for coordinating lower-level behaviors, rather than which classifiers own those behaviours' (*UML 2.2 Superstructure*, p. 311)). Rather, it is a specific action that is possible in a given 'activity' (in the sense of a business

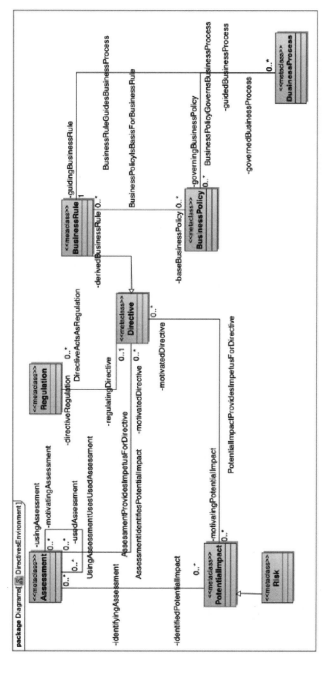

Figure 12.9   BMM model elements for regulatory compliance.

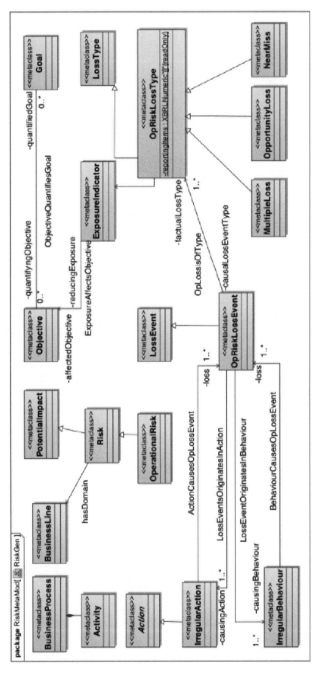

Figure 12.10   Operational risk metaclasses extending BMM.

process or subprocess) context. For example, the activity is to transfer money from account A to account B. In this activity, a fraudulent action would be to exchange B with another account ID, which means to change the parameters of the underlying transfer action. Note that having a fraudulent activity would mean having an intrinsically faulty business process – and it would be quite unusual to assume this in a risk management context. Therefore, the activity-related cause of an operational loss event should be modelled on the action, not on the activity level.

• Further consequence of a UML-conforming usage of the term 'activity' is that the cause of a loss event cannot be an activity in general but some 'malformed' action within an activity. This 'malformedness' can arise intentionally (fraud, theft, etc.) or unintentionally (inattentiveness, typing errors, etc).

The following principles guide the metamodel (see Figure 12.10):

1. A BMM business process is modelled as a composition of activities in the usual UML sense.

2. In order to capture alarm signals indicating(!) that the causes of a loss event are present, the abstract metaclass IrregularBehavior is introduced.

3. In order to capture actions causing loss events (see discussion above), the metaclass IrregularAction is introduced.

4. Both irregular actions or irregular behaviours are connected with an association to loss events.

5. The relationship between an action 'going wrong' (intentionally or not) or a system failure and a factual loss event should be modelled as a causal one-to-many association.

6. In addition, for risk analysis an association pointing in the opposite direction is needed – that is, linking observed loss events to potential causes in terms of action or system faults. This association is one-to-many again. In addition, since the exact causes in terms of system failures can be non-deterministically related to the loss events effected, this association should be non-navigable.

7. The operational risk loss events are subclasses of loss events. Loss events, since they result in factual losses, counteract at least one enterprise objective that quantifies some business goal according to BMM.

8. The loss events have types that describe the exact reporting implications of the loss. Types of loss in operational risk loss are write-downs, loss of recourse, restitution, legal liability, regulatory and compliance, loss or

damage to assets, see the BCBS operational risk document of 2007. (Write-downs: direct reduction in value of assets due to theft, fraud, unauthorized activity or market and credit losses arising as a result of operational events; Loss of Recourse: payments or disbursements made to incorrect parties and not recovered; Restitution: payments to clients of principal and/or interest by way of restitution, or the cost of any other form of compensation paid to clients; Legal Liability: judgements, settlements and other legal costs; Regulatory and Compliance (incl. Taxation Penalties): fines, or the direct cost of any other penalties, such as licence revocations; Loss of or Damage to Assets: direct reduction in value of physical assets, including certificates, due to some kind of accident (e.g. neglect, accident, fire, earthquake).)

9. To each loss event type, at least one exposure indicator is associated (or can be modelled as a property of the respective class). This exposure indicator impacts on a business objective – the association provides a link between the entire operational risk analysis and the business goal framework in BMM.

This metamodel should be seen in connection with the BMM modelling elements contained in our first metamodel in Figure 12.9 in order to appreciate the full compliance management impact. In particular, via the generalization of risk as a potential impact associated with assessments that provide the impetus for directives, we see that the risk analysis will contribute to business regulation settings and thus enable a full improvement cycle including risk mitigation and risk avoidance strategies. This is important, since the operational assessment and quantitative analysis of operational risk by itself contributes only to the *ends* elements in BMM (goals, as quantified by objectives). However, from an intelligent compliance management perspective, what is needed is a loop-back of these analyses to the *means* elements that ultimately allow us to define strategies, tactics and thus to derive specific directives governing business processes. In practice, the modification of directives in operational risk management often appears through so-called risk mitigation strategies.

12.6 Summary

This chapter presents an approach to intelligent regulatory compliance (IRC) in operational risk management. Our approach relies on specifications by the Object Management Group (OMG) fostering a holistic view of enterprise resources, processes and performance management under the umbrella of a business motivation model (BMM). An additional model detailing business vocabularies and business rules (SBVR) allows the definition of the directives, policies and rules of a regulatory framework in a formalized way. The practical implications of the approach, as of today, are limited to the modelling and planning phase of an intelligent management approach taking regulatory compliance fully into account. It is shown that the integrative enterprise modelling approach based on BMM and SBVR is sufficient to set up suitable management processes and business scorecards for enabling intelligent regulatory compliance.

The chapter concludes by showing how the next step of implementing IRC with appropriate software components could be reached, using the software engineering methodology advocated by the OMG, namely the model-driven architecture (MDA) together with UML. More specifically, an operational risk metamodel that combines BMM with related modelling elements from UML is provided. This is a basic setup that can be integrated with the model-driven enterprise engineering approach using SBVR. Such methodologies and supporting technologies are designed to help improve intelligent regulatory initiatives, a much needed capability of regulators, as discussed in Chapter 2.

The authors wish to thank the editors of the present volume for research initiatives related to intelligent regulatory compliance (IRC). Furthermore, the authors wish to thank the MUSING researchers Michele Nannipieri (MUSING partner Metaware) and Stefano Visinoni (MUSING partner Banca Monte dei Paschi di Siena) for their insight regarding the combination of IRC with failures/claims analyses in the context of near misses and opportunity losses that was implemented as one of the MUSING pilot applications.

References

Basel Committee on Banking Supervision (2003) Sound Practices for Management and Supervision of Operational Risks, www.bis.org/publ/bcbs96.pdf?noframes=1 (accessed 21 May 2010).

BRS, Business Rule Solutions LLC (2010) *BRS RuleSpeak® Practitioner's Kit*, 2001–2004, available from http://BRSolutions.com/p_rulespeak.php (accessed 21 May 2010).

EFIRM, European Financial Institutions Risk Managers Forum (2010) Typology of Operational Risk, http://www.efirm.org/Appendix1.pdf (accessed 21 May 2010).

Graml, T., Bracht, R. and Spies, M. (2008) Patterns of Business Rules to Enable Agile Business Processes, *Enterprise Information Systems*, 2, 4, pp. 385–402.

Hallé, S., Villemaire, R., Cherkaoui, O. and Ghandour, B. (2007) Model Checking Data-Aware Workflow Properties with CTL-FO+, *11th IEEE International Enterprise Distributed Object Computing Conference (EDOC)*, pp. 267–273.

Halpin, T. (2001) *Information Modeling and Relational Databases*, San Francisco,Morgan Kaufmann.

KnowGravity (2010) http://www.knowgravity.com/eng/ (accessed 21 May 2010).

MAGNA, Israel Securities Authority (2010) http://www.isa.gov.il/Default.aspx?Site= ENGLISH&ID=2860 (accessed 21 May 2010).

MDEE, Model Driven Enterprise Engineering™ (2010) KnowGravity, Inc., http://www. knowgravity.com/eng/value/mdee.htm (accessed 21 May 2010).

MUSING (2006) IST-FP6 27097, http://www.musing.eu (accessed 21 May 2010).

OMG (2010a) The Object Management Group, http://www.omg.org/ (accessed 21 May 2010).

OMG (2010b) The Governance, Risk Management and Compliance Global Rules Information Database, http://www.grcroundtable.org/grc-grid.htm (accessed 21 May 2010).

OMG, Business Motivation Model (BMM) (2008b) v1.0 – OMG Available Specification.

OMG, Semantics of Business Vocabulary and Business Rules (SBVR) (2008a) v1.0 – OMG Available Specification.

van der Aalst, W. and Pesic, M. (2006) DecSerFlow: Towards a Truly Declarative Service Flow Language, in *The Role of Business Processes in Service Oriented Architectures*, Dagstuhl, Internationales Begegnungs- und Forschungszentrum für Informatik (IBFI).

13

Democratisation of enterprise risk management

Paolo Lombardi, Salvatore Piscuoglio, Ron S. Kenett, Yossi Raanan and Markus Lankinen

13.1 Democratisation of advanced risk management services

Semantic-based technologies are proving very useful in the field of risk management. The general advantage of such technologies is twofold: on the one hand, they enable processing of large amounts of data, transforming such data into actionable information; on the other hand, they enable simultaneous usage of knowledge management techniques and statistical tools. From this standpoint, a true democratisation of the pervasive usage of risk management approaches and methodologies becomes possible. This is particularly relevant for industries other than financial services, where advanced risk management procedures are not necessarily imposed by regulatory supervisors, thereby opening up possibilities for a sustainable, voluntary usage of operational risk methodologies.

The operational risk management (OpR) methodologies are being taken to greater levels of complexity by the new needs of the financial services and insurance (FSI) industry in managing risk-adjusted performance of banks. As of today, such methodologies are being applied in industries other than FSI, mostly in large organisations or life-critical domains. In fact, although small- and

Operational Risk Management: A Practical Approach to Intelligent Data Analysis Edited by Ron S. Kenett
and Yossi Raanan © 2011 John Wiley & Sons, Ltd

medium-sized enterprises (SMEs) could greatly benefit from an active policy of OpR, a substantial barrier to adoption by SMEs is represented by the high cost of the aforementioned methodologies as they are currently known and applied. Another barrier to the effective usage of risk management methodologies is their perceived complexity and the need to have specially trained personnel doing risk management, usually at a cost that is considered by SMEs too high for its perceived value.

Conversely, one element of attractiveness, especially for SMEs, lies in the integrated management of operational risk and financial risk, a condition that enables enterprises to pair up a process view of their organisation with the financial dimension of their business, thereby allowing timely decision support in activities such as partner selection, risk-adjusted customer ranking and investment plans.

It is to be stressed that OpR methodologies can in principle be applied to:

- all industries;

- all enterprises (large or small).

Keeping that in mind, a possible representation of the value chain of risk management services powered by semantic-based technologies is provided in Figure 13.1.

Given the above scenario, the usage of semantic-based technologies can lower the OpR barriers to entry and develop a true democratisation of risk management services as described in the following sections. OpR techniques developed in the financial services can then be transferred to non-financial organisations, owing to the ontologisation of the discipline.

13.2   Semantic-based technologies and enterprise-wide risk management

As discussed above, democratised OpR services can provide access to a large number of SMEs in different industries and allow for effective implementation of risk management monitoring and mitigation strategies. Indirectly, this is linked to the business-process-oriented view of organisations. The bottom part of Figure 13.1 provides some details of specific advantages and technical enablers used to that aim. Semantic-based technologies can successfully and economically treat large quantities of data arising from the enterprise's processes. In fact, the central elements of the OpR value chain show that practical solutions can be achieved by means of semantic-based tools and techniques.

Semantic-based elements reduce most of the barriers to the adoption of OpR methodologies and processes, namely:

- event monitoring;

- data processing and handling;

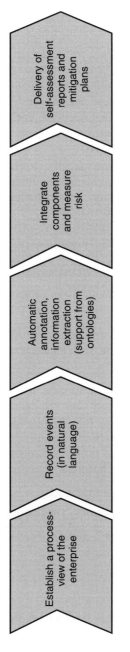

Establish a process-view of the enterprise

- This could be through a web based application-(SaaS).
- The process-based model opens up for further use of the schema (e.g., for correct sizing of the various offices of the SMEs responsibles for such processes).

Record events (in natural language)

- Both, machine-generated text (e.g.,log files from different sources) and manually inserted information in natural language (even in various languages) can be processed by the system.

Automatic annotation, information extraction (support from ontologies)

- The domain-ontology, specifically adapted to the involved SME, is the fundamental support for the annotation and information extraction software and routines.
- Human-readable, automatically-generated reports can be an intermediate product of extremely high value to the SMEs, in order to implement process improvement and risk reduction initiatives.

Integrate components and measure risk

- Conventional algorithms and models can be utilised here.
- Run-time integration of different risk categories can be performed (e.g., operational risk and financial risk)

Delivery of self-assessment reports and mitigation plans

- Human-readable, automatically generated reports are the ultimate product of the semantic-based system.

Figure 13.1   Risk services powered by semantic-based technologies.

- information extraction;

- information merging;

- risk analytics;

- report production.

All the above are fundamental elements whose management is effectively eased by semantic-based solutions. Moreover, if, on the one hand, semantic-based technologies can effectively democratise the usage of advanced methodologies, they also support runtime integration of different risk categories, for instance a combination of operational and financial risks, thereby opening up practical opportunities for enterprise-wide risk management (ERM).

A shortlist of the main elements constituting a semantically enabled ERM approach can be drawn directly from the outcomes of the MUSING four-year pan-European effort aimed at proving, with real-life cases, the industrial applicability of semantic-based technologies (MUSING, 2006).

Among such findings, the following are solutions that have been field tested and are particularly promising for industrialisation in different contexts:

- Risk-relevant information annotation and information extraction.

- Domain ontologies as the basis of automating time-consuming activities or production of reports without the manual intervention of analysts/domain experts.

- Bayesian integration of risk components (e.g. OpR with financial risk management).

- Methodologies for appreciation of the so-called 'near misses', that is events that did not actually generate an operational loss and yet damaged the organisation, for instance in terms of opportunity loss, highlighting areas to be closely monitored. These methodologies could reasonably be part of future regulations in OpR, contributing to assess further the robustness of the organisation's processes and stability of the business.

- Models for understanding the so-called 'multiple loss' events.

Overall, the potential impact on society is very relevant: there are several categories of SMEs that could immediately benefit from a sustainable, integrated approach to risk management:

- SMEs operating in the ICT field.

- Organisations that operate internationally and do not have the resources to finance a full-fledged risk management approach to their initiatives abroad.

- SMEs that pursue an evolutionary path in the line of advanced knowledge management, with particular reference to reorganisation of the company in a process-oriented view of their business.

Last but not least, those organisations that evolve in the above direction can also benefit from a privileged dialogue with their financial partners, inasmuch as they will find themselves speaking the banks' language in several relevant areas, such as risk control and risk mitigation.

13.3   An enterprise-wide risk management vision

Risk management is typically implemented by vertical approaches to individual risk categories. Even the most sophisticated organisations, such as financial institutions with an international presence, do not always achieve excellent results in managing individual risk categories. Moreover, the full benefits from an integrated risk management are far from being enjoyed (Power, 2004; Crouhy et al., 2006; Economist Intelligence Unit, 2009). The concept of ERM has received authoritative sustainment in the past few years (see Chapter 1 and Deloach, 2000; Lam, 2000; Bansal, 2003; Fouche, 2007). One interesting example of integration of financial risk management and OpR related to the field of electricity wholesale markets is documented in the work of Deng et al. (2009). However, not enough success stories have been brought to the attention of SMEs in order to have a chain-reaction impact in this area.

Organisations in general, and SMEs in particular, can experience particular benefits from the integration of risk categories, in that the tangible benefits arising from application of advanced risk management practices would considerably increase, and, at the same time, the cost barriers to entry would be reduced. Moreover, the effectiveness of a 'unique access point' to OpR is higher, especially in those situations of lower risk management culture. In fact, with an ERM approach, the sharing of information (and exploitation of such information) throughout the organisation becomes possible, allowing the application of risk management principles to the management practices of the organisation itself (see Chapter 1).

OpR is notoriously best treated by organisations that view themselves by focusing on processes. Pairing up events to processes provides a coherent map of the key risk indicators and permits the development of mitigation plans accordingly. However, OpR, even if treated with a pervasive approach, does not always give decision makers evidence on which aspects of the enterprise's processes need to be addressed in order to improve the financial viability of the organisation. At that point, adding the financial risk management dimension to OpR system provides the very valued ability to address effectively one of the urgencies of SMEs: accessing credit at the right price. That is why financial risk management is a discipline not just for financial organisations, but of cross-industry interest (Bansal, 2003; Power, 2007). As an example, in the case of organisations operating in the service provisioning business, they face both risk categories when making decisions about:

- enrolling a new customer;
- extending the service contract of an existing customer.

In fact, in both cases, providing a service to that customer means that the service provider assumes both an operational risk and a financial risk. The operational risk is created by the uncertainty in the distributions of both service demand and service costs; the financial risk is created by the uncertainty of receiving payment for the rendered services. Therefore, the two types of risk are present and a method is required to combine them into a single risk score, thereby streamlining the decision-making process and, at the same time, making the decision criteria objectively comparable across customers, across industries and in different timeframes.

The logical steps for a company that decides to adopt an ERM approach (with a self-assessment orientation) could be:

1. Schematise the organisation by means of a process-oriented model.

2. Personalise the ontology to the depth of the analysis that is required.

3. Define the risk types to be tackled.

4. Identify suitable KRIs.

5. Produce a 'risk assessment and action plan' report (and act upon it).

The main advantages for the enterprises that adopt an ERM approach could be summarized as follows:

- Ability to drive process management and improvement not simply by productivity or efficiency aspects, but also by risk considerations, which provide a more forward-looking perspective, with the correct economical view of the loss impact entailed with such processes.

- Minimisation of the potential losses and concentration of the resources on improving resilience of the most important business processes.

- Identification of the main correlations between OpR and financial risk management, with managerial indications on levers to be pulled in order to set proactively the 'risk appetite' of the organisation, together with an understanding of the most relevant entries to drive the business of the enterprise that, in a risk-centred view, should be directly connected to a correctly proportioned risk-taking attitude of every enterprise, especially those operating in international contexts.

- Development of a language that is common to all functions of the enterprise, thereby increasing involvement and trust among all employees directly involved in the process of maintaining up-to-date tools and databases at the foundation of the ERM system.

- Support the evolution of the organisation, also in terms of transparency towards external players, such as stakeholders, investors and the customers themselves. This has also a positive impact in terms of reduction in the costs of access to credit with financial institutions.

13.4   Integrated risk self-assessment: a service to attract customers

A semantic-based service can effectively be utilised to attract prospective customers. Typically, organisations that aim to internationalise their business are interested in developing a dialogue with financial institutions in order to access credit. Financial institutions, in turn, are always looking to acquire new customers, particularly those who operate internationally, because they require sophisticated services and often involve long-term relationships. They typically are interested in offering applications, for instance through their web sites, that generate customer leads that are actionable, for instance, through direct marketing initiatives.

A tool powered by semantic-based technologies that is interesting for organisations aiming to internationalise their business and appears sustainable for a financial institution looking for customer leads is a Web-based service, called Integrated Risk Check-up System for Enterprises (IRCSE), that combines the operational risk view with the credit risk analysis. The procedure is meant to be effective for a first check on an enterprise of any size and complexity, willing to take its first steps into an ERM approach. Such a solution demonstrates the added value of an integrated approach, which combines the OpR dimension and its typical process-based analysis with an analysis of financial aspects, from which credit merit considerations are derived.

The IRCSE solution, presented on the Web by a bank, and based on the semantic elements mentioned in the previous sections of this chapter, provides the following benefits to organisations undergoing self-assessment:

- A quantitative risk profile for the enterprise, covering both operational and credit risks.

- A preliminary action plan for improving the enterprise risk profile, with benefits from the credit merit point of view, from the insurance pricing point of view, as well as from the process optimisation point of view in general.

- A suggested spectrum of financial products and services, including term loan, support services for internationalisation, risk mitigation products, etc.

The IRCSE solution is described in Table 13.1, and is part of a plan that aims to take the bank–enterprise dialogue to a broader dimension, bringing the risk management perspective to a common ground for building a mutually transparent, successful partnership.

Table 13.1   Analysis of the IRCSE service.

Item	Value	Upside potential
Value proposition	To offer a risk self-assessment service dealing with operational risk and credit risk, delivered via the Web and founded on semantic technologies, with the aim of developing a more evolved dialogue with organisations, i.e. including analysis of the risk management dimension	The self-assessment check-up opens up for a pay-per-use self-assessment service for all enterprises, if it is to be delivered by a non-financial organisation
Target users	All enterprises (large to small), from all industries. Specifically, the service has been targeted to SMEs in the process of internationalising their business	
Main benefits for the end-users	• Facilitated access to credit, through a transparent reply • Risk management awareness • Action plan for improvement of the risk profile • Advantages to improve access prices to financial services in general, and bank assurance products in particular	

Main benefits for the financial institution delivering the service	• Database of prospect customers with no manual intervention or variable costs involved • High-level reputation/visibility; improved perceived value in transparency towards customers
Country coverage	Virtually all countries	Services set up in countries with strong inclination to internationalise local enterprises can be particularly successful
Language coverage	Multilingual, with English as underlying language	Localisation of the interfaces has been effectively tested also in Italian and German
Legal and regulatory issues	Privacy disclaimers always need to be localised
Customisability and technological engine	The service is based on a business process management platform (based on the BPEL standard executable language) that allows customisation of the service as it appears when delivered on the Web. The questionnaire to be filled out by the end-user is fully and dynamically adaptable

(continued overleaf)

Table 13.1 (*continued*)

Item	Value	Upside potential
Market potential	The rate of generation of records of prospective customers is initially directly related to the flow of unique visitors to the web site the service is placed in. Eventually, the service can contribute to increase traffic on the web site itself	
Barriers to entry on the solution	Semantic-based technologies are still a research topic. The underlying domain ontology, the information extraction components at the basis of the solution, and the Bayesian networks for risk integration are the most relevant technological and methodological barriers to entry	
Delivery channel	Internet	
Pricing	Free access in return for full privacy opt-in	The business model of IRCSE is founded on the concept of trading self-declared data with independent financial advice. This model can be changed to a fee for service, especially when the advisor is a non-financial institution

13.5   A real-life example in the telecommunications industry

Organisations can effectively rank customers in their portfolio, also by integrated risk analysis, accounting for factors such as operational risk and financial risk. From that perspective, an interesting test case of integrated risk management has been completed with the support of semantic-based technologies (MUSING, 2006). The case study was conducted by a leading telecommunications service provider that specialises in services for private branch exchanges (PBXs), covering:

- Installation of PBXs on the customers' premises.

- Remote diagnostics and repairs.

- Technicians' visits to the customers' sites, in order to perform diagnostics, visual inspections and repairs – when those repairs could be done on site. Repairs that could not be completed on site are referred to the service provider's engineering department for further investigation.

- Re-engineering existing telecommunications services to provide voice and data services to the clients, with a predefined quality of service and within a given cost structure. This is a quite sophisticated line of service, involving negotiation, on behalf of the customers, of various types of access lines and services and reconfiguring them in a way that best suited the customers' needs and budgets.

Besides the telecommunication component of the services, the business model involved service contracts with various payment schemas, as well as installation of PBXs, with annexed provisioning of the financial plan for them when the customers wanted it. For the customers, it was essentially a VNO leasing telecommunications equipment.

The service provider clearly needed to assess correctly the risks undertaken with the various customers. First, the operational risks that the service provider faced involved the frequency and the severity of breakdowns in telecommunications services to its customers. Frequency could be assessed from accumulated experience with the different equipment types that the service provider maintained. Severity, however, has a drastically different interpretation by different organisations. A typical example encountered is on service outages due to equipment breakdown. While in commercial organisations this was viewed as a problem requiring immediate solution – since lost communications meant lost revenue – often, in a public service organisation, the same incident provoked a quite different impact. Moreover, the service level agreement (SLA) required by these two organisations was drastically different in terms of response time, availability of replacement parts, delivery of substitute communications means, etc. These different requirements meant quite different costs which – to the service provider – meant problem severity. This meant that the nature of the malfunction or breakdown alone did not suffice for determination of the severity of the risk event and that severity had to be assessed by different, more fine-tuned means.

Another issue that the service provider faced was that of the commercial viability of its clients. Whereas in service contracts its exposure was reasonable – all at stake in case of non-payment were technicians' work hours and perhaps some replacement parts – in those cases where the service provider actually supplied the PBX and its ancillary equipment, the financial loss in cases of insolvency was much higher.

All in all, what was really needed was a tool for assessing the combined effect of financial and operational risk of the various business transactions. However, even assuming it could develop risk indicators for both operational risks and financial risks, which it did not, the service provider would not know how to treat a customer with, say, an OpR score of 0.5 (on a scale of 0 to 1, 1 being the highest risk) and a financial risk score of 0.2 (on the same 0 to 1 scale), versus, for example, another customer with an OpR score of 0.3 and a financial risk score of 0.7. Therefore, a combined risk index was designed and tested using the methodologies described in Chapter 8. Such a combined score enabled a more coherent, compact approach to business decision making by putting into a single measure a 'total customer risk', as assessed by the company, thereby enabling a more balanced appreciation of the various customers.

13.6 Summary

Semantic-based solutions (e.g. processes supported by machine learning techniques, ontology-driven approaches, natural language processing, qualitative–quantitative information extraction and its merging) enable the automation of several tasks, such as information extraction from large amounts of textual information, guidance of non-technical users in complex risk management activities, and integration of risk categories that are relevant for a broad spectrum of SMEs (from various industries), such as operational risk and financial risk. This is particularly relevant for organisations that plan to internationalise their business.

The semantic-based tools and solutions that have been field tested indicate that Web-based risk management services can be effective in a substantial part of a value chain of risk knowledge services, opening up the democratisation of operational risk practices and, more in general, of enterprise-wide risk management.

A process-oriented platform of risk management services open for the end-user to configure its required service level is indeed not only feasible, but also sustainable for both the provider and the end-user. The social networking phenomenon will positively influence the spreading of self-assessment solutions, thereby opening up even more business opportunities in this field.

References

Bansal, P. (2003) Enterprise-wide risk management, Reprinted from *The Banker*, February.

Crouhy, M., Dalai, D. and Mark, R. (2006) *The Essentials of Risk Management*, McGraw-Hill.

Deloach, J. (2000) *Enterprise-wide Risk Management – Strategies for linking risk and opportunity*, Prentice Hall.

Deng, S., Meliopoulos, S. and Oren, S. (2009) Integrated Financial and Operational Risk Management in Restructured Electricity Markets, Power Systems Engineering Research Center (PSERC) Publication.

Economist Intelligence Unit (2009) After the Storm – A New Era for Risk Management in Financial Services, White Paper (2009).

Fouche, C. (2007) Enterprise Wide Risk Management: Explained, http://christelfouche .com/blog/ewrm-explained (accessed 21 May 2010).

Lam, J. (2003) *Enterprise Risk Management: From Incentives to Controls*, Wiley Finance.

MUSING (2006) IST-FP6 27097, http://www.musing.eu (accessed 21 May 2010).

Power, M. (2004) *The Risk Management of Everything – Rethinking the Politics of Uncertainty*, Demos.

Power, M. (2007) *Organized Uncertainty – Designing a World of Risk Management*, Oxford University Press.

14

Operational risks, quality, accidents and incidents

Ron S. Kenett and Yossi Raanan

14.1 The convergence of risk and quality management

Quality, as a subject, has been developed by Shewhart, Deming, Feigenbaum and Juran and others. Joseph M. Juran's contributions include structured process improvement and quality by design methods aimed at lowering operating costs and increasing revenues and profits. Quality improvement, quality planning and quality control initiatives are known as the Juran Quality Trilogy and form the basis of total quality management and Six Sigma (Juran, 1986, 1989). Godfrey and Kenett (2007) review Juran's contributions and emphasize the need to integrate information from different sources in order to achieve better decision making. Here, we focus on the integration of quality and risk as an approach leading to increased insights and knowledge, and thereby better management decisions. A risk and quality convergence supports consumers and firms seeking to prevent and control risks and their consequential effects. While, as discussed in Chapters 1 and 2, risks define the consequences of an adverse event, poor quality of product and service design represents an 'unrealized' expectation. Poor quality of process is reflected by the amount of rework, scrap or recalls. In this sense, both quality and risk deal with adverse consequences and are subjected to probabilities that define their occurrence. Further, both take into considerations individuals, firms and society at large. The issues relating to who bears

Operational Risk Management: A Practical Approach to Intelligent Data Analysis Edited by Ron S. Kenett and Yossi Raanan © 2011 John Wiley & Sons, Ltd

the risk if it occurs, as well as who bears the responsibility for producing a non-quality product is essential and important in both cases. Risk and poor quality can result from many causes, both internally induced, such as inappropriate training and contradicting organizational goals, or occurring externally, for example by water quality or economic conditions. Their consequences, however, may be diverse, affecting various parties. When risk is internally induced, it may be due to low-capability operations, faulty operations, human error and failures or misjudgement. See for example an analysis of aircraft accidents and aircraft incidents (Kenett and Salini, 2008). Similarly, when risk is endogenous, it results from systemic effects. Preventive efforts and insurance can be used to mitigate such consequential effects. In general, a combined definition of risk and quality involves the following factors:

- **Consequences**, bourn individually or collectively, by the persons responsible for the adverse risk or quality event, or by others. This represents a future event.

- **Probabilities and their distribution**, assumed known, partly known or not known, consisting of random (systemic) recurrent, persistent or rare events. These represent past experience.

- **Detectability**, reflecting our ability to identify the risk. Low detectability obviously requires more aggressive mitigation strategies.

- **Individual preferences and risk attitudes**, representing the costs, subjective and psychological effects and needs, and a personal valuation (price) of the needs and their associated consequences.

- **Collective and shared effects**, including effects on society at large (or risk externalities) and the manner in which these risks are negotiated and agreed by the parties involved – either through a negotiated exchange or through a market mechanism.

- **Pricing of risks and quality**, usually based on an exchange within an organization (as negotiated in typical industrial contracts) or occurring in specific markets (financial or otherwise) where such risks are exchanged.

For more on these definitions see Tapiero (2004), Haimes (2009) and Kenett and Tapiero (2010).

These definitions imply that risk and quality share many common concerns. Quality is in many cases derived from the risks embedded in products or processes, reflecting consumer and operational concerns for reliability design, maintenance and prevention, statistical quality control and statistical process control and the management of variability. Similarly, the many tools used in managing risks seek to define and maintain the quality performance of organizations, their products, services and processes. Both risks and quality are relevant to a broad number of fields, each providing a different approach to their measurement, their valuation and their management which are motivated by psychological, operational, business and financial needs, as well as social needs and norms. Both deal with problems that result from uncertainty and adverse consequences which may be predictable or unpredictable, consequential or not, and express a like or a dislike for the events and consequences induced.

Risk and quality are thus intimately related, while each has, in some specific contexts, its own particularities. When quality is measured by its value added, and this value is uncertain or intangible, uncertainty and risks have an appreciable effect on how we measure and manage quality. In this sense, both risk and quality are measured by 'money'. For example, a consumer may not be able to observe directly and clearly the attributes of a product. And, if and when he/she eventually does so, this information might not be always fully known, nor be true. Misinformation through false advertising, unfortunate acquisition of faulty products and product defects have a 'money effect' which is sustained by the parties involved. By the same token, poor consumer experience in products and services can have important financial consequences for firms that are subjected to regulatory, political and social pressures. Poor quality, in this sense, is a risk that firms should assess, seek to value and price. Finally, both have a direct effect on value-added offerings and are a function of the presumed attitudes towards risk and quality by consumers.

The approach used to manage and share risks, from businesses-to-consumer, consumer-to-consumer and business-to-business transactions, is immensely important. It represents essential facets of both the process of risk and quality management. Warranty contracts, service contracts, liability laws and statistical quality control are some of the means available to manage these risks and thereby quality (see Tapiero, 2004; Kogan and Tapiero, 2007). Conversely, managing risks through preventive measures, Six Sigma and related techniques improves the prospects of quality as well. Of course, each situation may have its own particularities and therefore may be treated in a specific and potentially different manner. For example, environmental pollution and related issues have both risk and quality dimensions that may be treated differently than, say, the quality and the risks of services (Tapiero, 2004).

As another example, consider the definition of service quality. A gas station provides several services beyond the supply (frequently at a regulated price) of fuel. Hotels provide a room and various associated services. As a result, the quality of service may be defined mostly in terms of intangibles, often subjective and therefore difficult to define, unless their satisfaction can be specified by a risk (probability) event. Unlike quality in manufacturing, the quality of services depends on both the 'service provider' and the 'serviced customer' with their associated risks. Poor service is usually reflected by customer dissatisfaction and eventual churn. Service delivery needs to be consistent, providing the right service, every time. Comparable notions in industry are addressed by considering machine breakdowns or improperly performed functions. Furthermore, the quality of service and its measurement are dependent and at times subjective. A service provider who is inspected might improve the quality of service delivery. A sense of lack of controls might result in poor service delivery. Such behaviour introduces a natural bias in the measurement of service efficiency and its quality, which can benefit from a probabilistic and risk-based approach. These specific characteristics have an important impact on the manner in which we conceive and manage both the quality of service and its associated risks, and their consequences to individuals, firms and society.

The following section provides a conceptual framework integrating quality and risk. In particular it emphasizes the four quadrants proposed by Taleb as a framework to systematize a quality–risk convergence (Taleb, 2007, 2008a, 2008b). Such a framework has both managerial and technical connotations that do not negate the fact that risk is complex, at times unpredictable and at times of extraordinary consequences. A convergence of quality and risk enriches two important and creative areas of research and practice, augments transparency and provides a measurable value of quality and a better assessment of what we mean when we define quality. Quoting Robert Galvin, former Chairman of the Board of Motorola Inc.:

> Perfect quality, perfect delivery, perfect reliability, perfect service – these are achievable. ... The quality system that will be embraced by any particular organization who takes the subject very seriously will aim for those goals and be measurable by the appropriate and dedicated use of the statistical systems that are now readily available.
> (From the Foreword by R. Galvin to the *Encyclopedia of Statistics in Quality and Reliability*, Ruggeri *et al.*, 2007)

In general, our goal is to better generate knowledge based on information and data as formulated in the introduction to Chapter 1 (see also Kenett, 2008; Kenett and Shmueli, 2009). We proceed to present Taleb's four quadrants in the context of the quality and risk management integration.

14.2   Risks and the Taleb quadrants

Important consequential risks are typically unpredictable and rare. While predictable risks may be prevented, unpredictable risks test our resilience and our ability to respond. Based on these premises, consider Nassim Taleb's metaphor: 'A Turkey fed for 1000 days, every day, confirms that the human race cares about its welfare with increased statistical significance. On the 1001st day, the turkey has a thanksgiving surprise, its total demise.' Similarly, the presumption that those good times were to last for ever can be remarkably irresponsible from a risk management viewpoint. The financial meltdown of 2008 may attest to such risks with an aggregate fate of close to 1000 financial institutions (including busts such as FNMA, Bear Stearns, Northern Rock, Lehman Brothers, etc.) that lost over $1 trillion on a single error, more than was ever earned in the history of banking. For more on turkeys, 'black swan' rare events and the economic meltdown, see Taleb (2007) and Kenett (2009). In this sense, Taleb's contribution of 'black swan' risks has contributed to a greater focus of risk analysis and its management on the rare and unpredictable spectrum, compared with the 'normal' risks that the statistical and financial risk analyst has traditionally addressed. To a large extent, similar problems have confronted the management of quality and reliability – from control in the early twentieth century to risk prevention in the latter part of that century and to robustness and quality by design (see Kenett and Zacks, 1998; Meeker and Escobar, 2004; Nasr, 2007; Kenett et al., 2008; Kenett and Kenett, 2008). The latter approach, robustness, emphasizes greater sensitivity to the mis-assumptions that underlie, necessarily, traditional models used to manage quality.

To confront this evolving risk reality, Taleb has suggested a mapping of randomness and decision making into four quadrants representing two classes of randomness and decisions. The type of decisions referred to as 'simple' or 'binary' lead to decisions such as 'very true or very false', 'matters or does not matter'. By the same token, statistical tests in the control of quality may state: 'A product is fit for use or the product is defective.' Statements of the type 'true' or 'false' can then be stated with some confidence interval. A second type of decision is more complex, invoking both its likelihood of occurrence and its consequences.

By the same token, two layers of randomness, very distinct qualitatively and quantitatively, are suggested by Taleb. A first layer is based on 'forecastable events', implied in finite variance (and thus thin-tailed probability distributions) and a second based on 'unforecastable events', defined by probability distributions with fat tails. In the first domain, exceptions occur without significant consequences since they are predictable and therefore preventable (or diversified in financial terms). The traditional random walk, converging to Gaussian–Poisson processes, provides such an example. In the second domain, large consequential events are experienced which are more difficult to predict. 'Fractals' and infinite variance (Pareto-stable and chaotic) models provide such examples (see Mandelbrot, 1982). These models presume that random processes in nature, or in financial markets, do not necessarily follow a Gaussian distribution (Taleb, 2008b; Chichilnisky, 2010). Thus to relieve the constraining assumption of such assumptions, weaker forms of underlying risk probability processes are suggested, such as Lévy stable distributions (or Levy processes) that have both leptokurtic distributions with potentially infinite variance. For example, in commodity prices, Mandelbrot found that cotton prices followed a Lévy stable distribution with parameter α equal to 1.7 rather than 2, as is the case in a Gaussian distribution. 'Stable' distributions have the property that the sum of many instances of a random variable follows the same distribution and therefore aggregates have the same distribution of their individual events.

These two dimensions form a map with four quadrants (Table 14.1), each quadrant appealing to its own methods to deal with the challenges that each quadrant is raising. For example, in the first quadrant, simple binary decisions, in cases of thin-tailed distributions with predictable events, lend themselves to effective statistical analysis which we tend to emphasize because of our ability to treat such problems successfully. Most real problems, however, do not fall in this quadrant. The second quadrant consists of simple decisions, confronted by 'heavy-tailed distributions'. Currently, important efforts are devoted to problems of this sort in finance which assume that financial markets are incomplete

Table 14.1   The four quadrants of N. Taleb (adapted from Taleb, 2008).

Domain/application	Simple decisions	Complex decisions
Thin-tailed 'Gaussian–Poisson' distributions	I. Classical statistics	II. Complex statistics
Heavy-tailed or unknown 'fractal' distributions	III. Complex statistics	IV. Extreme fragility ('limits of statistics')

or the underlying 'randomness' has a 'leptokurtic bias' combined with extreme volatilities in which financial markets seem to react chaotically, augmenting asset price volatility. The third quadrant deals with complex decisions in thin-tailed distributions where statistical methods work surprisingly well. In such cases, Monte Carlo techniques, appropriately designed, have provided an efficient means to investigate and solve problems related to the third quadrant. In this sense, while the first three quadrants may lead to complex risk analysis problems, these problems may adopt both modelling and computational techniques which can be used to remove and manage some of their associated complexity and uncertainties.

The risk challenge lies in the fourth quadrant, combining complex decisions with heavy-tailed (unpredictable events) distributions. Such situations occur when confronted with a black swan (although rare, existing nonetheless). These problems are important equally in the control and management of quality, reliability, safety and all matters where risks are prevalent. In particular, in large network-based and complex dependent systems, the interactive behaviour of these systems may lead to fractal models characterized by both unpredictability and catastrophic consequences (Meeker and Escobar, 2004).

Modern industrial organizations, in manufacturing and services, are characterized by increased networking and dependencies, by a growth of complexity, increased competitive pressures and rising customer expectations. While these expectations may mean seeking greater predictability, complexity of systems and products may not be able to meet the demand for such predictability. A growth of complexity and interdependence might overwhelm our physical capacity to circumvent such complexity. Such a phenomenon is embedded in Ashby's second law of cybernetics, the law of requisite variety: 'For a system to be stable, the number of states of its control mechanism must be greater than or equal to the number of states in the system being controlled' (Ashby, 1958). The next section investigates the role of management maturity on the ability of an organization to manage risk and quality efficiently with effective control mechanisms.

14.3   The quality ladder

Quality, as practised traditionally, is mostly focused on problem detection and correction. Problem correction, while essential, only serves to remove defects embedded in the product by development and production process. When properly organized for continuous improvement and quality management, organizations focus also on problem prevention in order to improve the quality of the product and their competitive position. Continuous improvement, as a company-wide management strategy, is a relatively recent approach. For thousands of years improvement and innovation were slow. New scientific breakthroughs often occurred by chance with an 'intelligent observer' in the right place at the right time. As significant events occurred, someone asked why, and, after some experimentation, began to understand the cause and effect uncovered by that observation. The 1895 discovery of X-rays by W.C. Roentgen is a classic example. By the late 1800s, an approach to centrally planned innovation and improvement was beginning to appear. Thomas Edison built his laboratory in 1887 and conducted thousands of experiments. Throughout the twentieth century several industrial, academic and government laboratories, often employing tens of thousands of researchers, were established. The art and science of experimental design became widely used to drive improvements in products and processes, and in developing entirely new products and services that are far more responsive to customers. In the latter half of the twentieth century, another phenomenon took place, first in Japan and then quickly in other parts of the world. Large numbers of employees in organizations were taught the basics of the scientific method and were given a set of tools to make improvements in their part of the company. They were empowered to introduce changes in processes and products in order to achieve improved production and product performance. Total quality management and Six Sigma are examples of this approach (Godfrey and Kenett, 2007).

Problem prevention (just as risk management) has two aspects: (1) preventing recurrence of existing problems, and (2) preventing introduction of new problems. In such cases, problem prevention results in quality improvement of two types: reactive (driven by problems) and proactive (driven by the desire to improve quality and efficiencies). Reactive quality improvement is the process of understanding a specific quality defect, fixing the product, and identifying and eliminating the root cause to prevent recurrence. Proactive quality improvement is the continual cycle of identifying opportunities and implementing changes throughout the product realization process, which results in fundamental improvements in the level of process efficiency and product quality. A reactive causal analysis of a quality defect will sometimes trigger a change in a process, resulting in a proactive quality improvement that reduces defect levels.

As mentioned in Section 14.1, Juran made significant contributions to the establishment of proper structures for quality improvement (Juran, 1986, 1989). Juran's Quality Trilogy is possibly the most complete representation of managing for quality ever devised. Juran developed the management of quality using an analogy that all managers easily understand – managing budgets. Financial management is carried out by three fundamental managerial processes: budget planning, usually on an annual basis, financial control reports and cost reduction initiatives. Managing quality is based on the same three fundamental processes of planning, control and improvement. The Juran trilogy consists of three phases, or types of activity:

1. Quality planning: The process for designing products, services and processes to meet new breakthrough goals.

2. Quality control: The process for meeting goals during operations.

3. Quality improvement: The process for creating breakthroughs to unprecedented levels of performance.

To implement the trilogy properly in products, services and processes, one must understand that the trilogy is three dimensional and limitless, resulting in an endless improvement process. For example, quality planning of a new market offering will consist of quality planning for products, services, processes, suppliers and distribution partners with an impact on the delivery of the new offering. On another layer, this phase must also plan for the support and maintainability of the offering. On yet another layer, the planning phase must account for the design and integration of data collection, control, improvement processes, people and technologies. Finally, the planning phase must design the evaluation of the planning phase itself. In this simple example, we see four layers of quality planning. This phase will typically be iterated and improved with every cycle and within each cycle. The same is true of all other phases. Such an approach, however, is incremental, and based on a recurrent and predictable process fed by a (Bayesian) experience that arises from 'learning by doing'. Striving for zero defects, perpetual improvement and prevention, appropriately combined with robust design and risk recovery, provides an avenue to meet the challenges of Taleb's fourth quadrant described in Section 14.2.

Management teams in global supply chains, on all five continents, are striving to satisfy and delight their customers while simultaneously improving efficiencies and cutting costs. In tackling this complex management challenge, an increasing number of organizations have proved that the apparent conflict between high productivity and high quality can be resolved through improvements in work processes and quality of designs. Different approaches to the management of organizations have been summarized and classified using a four-step quality ladder (see Kenett and Zacks, 1998; Kenett et al., 2008). The four management approaches are (1) fire fighting, (2) inspection, (3) process control and (4) quality by design and strategic management. In parallel with the management approach, the quality ladder lists quantitative techniques that match the management sophistication level (see Figure 14.1). This matching is similar in scope to Taleb's four quadrants. In this case, however, the scope is to match management maturity with the statistical techniques that can be effectively used by the organization.

Managers applying reactive fire fighting can gain from basic statistical thinking. The challenge is to get these managers to see the value of evolving their organization from a state of data accumulation to data analysis and proactive actions, turning numbers into information and knowledge. Managers who attempt

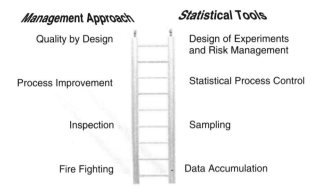

Figure 14.1   The quality ladder.

to contain quality and inefficiency problems through inspection and 100% control can increase efficiency by using sampling techniques. Their approach is more proactive than fire fighting but the focus on end products, *post factum*, can be very expensive. Sampling inspection can reduce these costs provided proper statistical analysis is used in order to establish sample sizes for effective inspections. The decision of what to test, when and where, should be assessed statistically so that the performance of the approach is known and adequate. More proactive managers, who invest in process control and process improvement, can take full advantage of control charts and process control procedures. Process improvements and risk prevention affect 'how things are done', thereby affecting both cost and quality in a positive way. At the top of the quality ladder is the quality by design approach where upfront investments are secured to run experiments designed to optimize product and process performance. At such levels of management sophistication, robust experimental designs are run and risk management and reliability engineering are performed routinely. Moreover, risk estimates are compared with field returns data to monitor the actual performance of products and improve the organization's predictive and planning capability.

The Statistical Efficiency Conjecture discussed in Kenett *et al.* (2008) states that organizations increasing the management sophistication of their management system, moving from fire fighting to quality by design, enjoy increased benefits and significant improvements with higher returns on investments. In this sense, recognizing the consequential effects of risks is not a theoretical exercise but a rational and economic consideration. The move up the quality ladder is pursued by management teams in different industries and in different ways. For example, electronic systems design, mechanical parts manufacturing, system assembly software-based services and chemical processes use different approaches embedded in their experience and traditions.

A particular industry where such initiatives are driven by regulators and industrial best practices is the pharmaceutical industry. In August 2002, the Food and Drug Administration (FDA) launched the pharmaceutical current Good

Manufacturing Practices (cGMP) for the twenty-first century initiative. In that announcement, the FDA explained the agency's intent to integrate quality systems and risk management approaches into existing quality programmes with the goal of encouraging industry to adopt modern and innovative manufacturing technologies. The cGMP initiative was spurred by the fact that since 1978, when the last major revision of the cGMP regulations was published, there have been many advances in design and manufacturing technologies and in the understanding of quality systems. This initiative created several international guidance documents that operationalize this new vision of ensuring product quality through 'a harmonized pharmaceutical quality system applicable across the life cycle of the product emphasizing an integrated approach to quality risk management and science'. This new approach is encouraging the implementation of quality by design and hence, de facto, encouraging the pharmaceutical industry to move up the quality ladder (see Kenett and Kenett, 2008; Nasr, 2007). The Capability Maturity Model Integration (CMMI) invokes the same principles for the software and systems development industry (Kenett and Baker, 2010). The next section discusses risks, accidents and incidents which are further discussed in Section 14.5 in the context of the oil and gas industry. The examples from this industry can be generalized to other application areas such as health care, transportation and air travel, or the building industry (Amalberti, 2001; Andersson and Menckel, 1995)

14.4   Risks, accidents and incidents

An accident is defined as an unexpected, unplanned or unwanted event that may cause damage, injury or illness to people. For example, an accident may interrupt the production and flow of a work process. Others define accidents as any undesired circumstances which give rise to ill health or injury; damage to property, plant, products or the environment; production losses; or increased liabilities (Johnsen and Everson, 2003).

Accidents occur in all types of daily activities as well as production and service activities. The accident causation process is complex and varies from one type of activity to another. Accidents are caused by the transfer of an excessive amount of energy from objects or substances. Occupational accidents are accidents which have consequences on the working process and workplace and may cause mortality. However, they often have no potential to cause fatalities outside the immediate area of the accident.

Incidents are undesired events with specific consequences and may contain near misses with the potential to cause accidents. The term 'potential' is particularly important in the investigation of incidents which have the potential to cause severe harm even if the actual harm caused was trivial. Muermann and Oktem (2002) define 'near miss' as an event, a sequence of events, or an observation of unusual occurrences that possesses the potential of improving a system's operability by reducing the risk of upsets, some of which could eventually cause serious damage. A near miss can be thought of as an event that could have resulted in a pecuniary loss, if it had been completely fulfilled. In that sense, the event was near to producing a real loss, but did not develop up to the level of its completion.

In defining a near miss it is important to include both negative and positive impacts. The near miss should be viewed from various aspects, such as operational disturbance and improvement opportunities capturing both events and observations. Therefore a near miss can be defined as an opportunity to improve environment, health and safety practice based on a condition, or an incident with the potential for more serious consequence (Rahim and Kenett, 2008a, 2008b). Chapter 10 provides a comprehensive treatment of quantifying non-recorded losses in the banking industry. This case study can be extended to other industries having similar databases. Computer and communication near misses are, in a sense, relatively easy to predict since most computer devices automatically collect and store maintenance data, error logs and historical data. Moreover, in many systems providing telecommunications and computing services, a centralized monitoring and control system is installed in parallel with the operational system. A study of the individual maintenance and log files of each piece of equipment as one type of input, with the central management and control database serving as another input, can yield significant insights into the extent and type of near misses and their impact. With this knowledge, proactive preventive maintenance activities such as changing configurations, components or procedures can result in better quality performance and reliability. When this knowledge is further augmented with semantic-based knowledge derived from customer relationship management (CRM) systems, even better insights can be gained for identifying effective steps towards better overall performance.

One important aspect of risks and near misses is their impact on the work environment. In general, however, there is no clear tracking of the cost and burden of work accidents. The burden of accidents is considered to be vast and takes a large share of resources in companies as well as in national social insurance funds. The process of estimating these costs requires comprehensive information of different elements, such as the causes, the mechanism and the consequences of such accidents. The method requires that these elements are reported, investigated and estimated correctly.

Companies need to have an efficient incident/accident reporting system, which identifies the main risk factors facing their operations, activities and sustainability and facilitates an effective accident investigation and learning process. Since many accidents may trigger civil and sometimes even criminal proceedings, the reporting itself becomes a risk event in many cases, thus severely limiting the data collection process (Attwood *et al.*, 2006). A prime example occurs in the health industry. While it seems that this is an industry that can significantly benefit from monitoring near misses, the fear of criminal liability and of tort awards has, at times, curtailed the use of morbidity and mortality meetings in hospitals. Such meetings should be held after an incident (near miss) or accident in order to investigate its causes. It is a bitter irony that hospitals, in an attempt to reduce the risk of paying out large settlements to injured patients or their families (and also of damaging their reputation), are sometimes adopting an attitude that actually increases the risk to their future patients. Other examples include education (Raanan, 2009) and air traffic control (NASA, 2002; Nazeri et al., 2008; Kenett and Salini, 2008a, 2008b).

Increased prevention and process improvement initiatives are highly dependent on the process of incident follow-up and investigation, in order to ensure that the underlying as well as immediate causes of accidents and incidents are disclosed and understood, taking full account of all causation factors such as human, technological, environmental and organizational factors. Chapter 9 discussed the application of association rules data mining techniques to such data. For a general analysis of cause and effect, see McKinnon (2000) and Kenett (2007).

Accident/incident investigation is an important qualitative/quantitative approach to understanding and managing industrial safety. It is a process for analysing cumulative information from both internal and external events, through analysis of available data accumulated from reporting systems. Use of appropriate techniques for investigations is essential for recommending necessary modifications and changes in order to prevent a recurrence of the incident or similar incidents in the future (see Schofield, 1998; Rahim and Kenett, 2008a, 2008b).

Measuring the burden of accidents requires a comprehensive understanding of the complex process of accident causation. Measuring safety requires an overview of dominant scenarios, their barriers and finally the resulting central events (Swuste, 2007; Rahim and Kenett 2009, Rahim, 2010). The computation and presentation of risk scores were the topics of Chapter 7, Chapter 8 and Chapter 11. The next section provides an extensive discussion of these topics in the oil and gas industry.

14.5   Operational risks in the oil and gas industry

To expand further on the concepts of risks, accidents and incidents we focus next on an application of operational risk management to the energy sector. The oil and gas industry provides energy and essential chemicals for our transport, industry and homes, and contributes to national economies with valuable tax revenues. The relationship between branches, industries and their population of workers and their environment is dynamic. Exposures to risks and injuries vary among branches and industries and change over time with the workplace physical environments (Bea, 2001; Aven and Vinnem, 2007).

The economic growth of nations is often followed by an increase in the demand for oil and gas products needed to run industries, transport, etc. The energy sector, especially the oil and gas sector, faces a huge demand for its products. The current high price of crude oil facilitates access to new international fields, licences and opportunities. The exploration of oil and gas has become much more challenging, the largest companies dominating and controlling the technology and facilities. Different groups and industries react differently to risks and injuries. The oil and gas industry carries very high risks in its exploration, production, processing and transportation projects. These risks vary across different operations and activities and incorporate both internal and external factors. In most offshore oil and gas activities, risk assessments are formulated in such a way that many probabilities are interpreted classically, that is as relative frequencies rather than as degrees of belief (Schofield, 1998). The elements contributing to the internal risk factors can be characterized as design, access, exposure, health, working hours, working conditions, etc. The elements contributing to the external factors include political instability, unstable oil and gas markets and constant changes in government regulations and guidelines.

The state regulator usually plays a major role in shaping the safety level and risk acceptance criteria in any industry. In most countries, the level of safety and risks is assessed in accordance with the legislation (see Table 14.2). National legislations encourage and promote a health, environment and safety culture comprising all activity areas.

The regulations in the oil and gas industry make it compulsory to report incidents to the national authorities. The companies in that industry are also required to submit an annual report concerning load-bearing structures, summarizing operational experience and inspection findings. The reporting is necessarily based on a classification derived form an implicit or explicit ontology of operational risks, as discussed in Chapter 3. Such an ontology can be used to analyse semantic data such as text and video captures (see Chapter 4). Based on this data, several types of statistical analysis are conducted to determine the safety level of specific companies.

The information from oil companies is compiled and analysed by regulators and the statistics and trends are presented for learning purposes, enforcement of legislation and prevention actions. Trend analysis is one type of analysis used

Table 14.2   State and industry norms and controls.

	Prescribing norms	Safety control
State regulatory and safety management	Safety requirements: • Legislation • Administrative decisions • Guidelines • Standards	State safety control and inspections
Industry safety management	Choice within the framework Standards Best practice	Internal safety control Third-party safety control

to determine the increase or decrease in safety levels for a particular industry, sector or geographical area.

Chapter 12 dealt with the new technology used in intelligent regulation of the financial industry. Such technology-based intelligent regulatory systems are becoming critical in order to meet the demands of an increasingly sophisticated financial industry and modern telecommunications infrastructures.

Regulatory authorities continue to develop new inspection and auditing concepts, conducting incident and accident investigations and analyses (see the Foreword to this book by Dr Marco Moscadelli from the Bank of Italy). Regulations in the petroleum activities are normative and contain functional requirements. This means that the regulations state the level of safety that must be achieved, but not how to achieve it, and what has to be achieved rather than providing concrete solutions.

The terms 'accident' and 'incident' can be interchangeable and, as mentioned above, include 'near miss' situations. Near misses occur more frequently than compulsory reportable events, like accidents. Therefore, monitoring near misses can identify trends where barriers are challenged and the safety is threatened prior to an accident (see also Chapter 10). In order to understand the impact of a near miss, and near miss management, on safety promotion and reducing the potential of accidents, we need to understand better the near miss mechanism and the importance of near misses in implementing a safety management programme.

From a safety perspective in the chemical, oil and gas industry, a near miss is an event which has the potential for an accident involving property damage, environmental impact or loss of life, or an operational interruption. This event or combination of events could have resulted in negative consequences if circumstances had been slightly different. In other words, a near miss provides a valuable opportunity to learn from deficiencies or other management system weaknesses that could help avoid possible future accidents (CSB, 2007).

Near misses include process-related disturbances, spills, property damage, injuries to employees and business interruptions, as well as natural disasters and terrorist events, which occur less frequently. Near misses cover a wide scope of conditions and events. These include:

- Unsafe conditions.

- Unsafe acts.

- Unsafe behaviour.

- Unsafe equipment.

- Unsafe use of equipment.

- Minor accidents that had the potential to be more serious.

- Minor injuries that had the potential to be more severe.

- Events where injury could have occurred but did not.

- Events with property damage.

- Events where a safety barrier is challenged.

- Events where business opportunities are missed.

- Events with potential environmental damage.

By identifying, analysing and dealing with near misses, adverse catastrophic events in chemical plants can be significantly reduced (Jones *et al.*, 1999; Phimister *et al.*, 2003).

Incident data is under-reported and subjected to self-reporting bias (Aven, 2003; Yashchin, 2007; Nazeri *et al.*, 2008, Kenett and Salini, 2008a, 2008b). Incident reporting databases may also be biased to individual tendencies to over- or under-report certain types of events (van der Schaaf and Kanse, 2004). The organizations' understanding and interpretation of how incidents and near misses influence how it collects the data on issues related to safety and performance. Other studies stated that focusing on data for near misses may add noticeably more value to quality improvement than a sole focus on adverse events (Barach and Small, 2000). Personal injury and occupational illness data is collected and maintained for all operators in the North Sea oil and gas industry and reported to various databases or national safety authority databases. This is a requirement containing basic information on the event, the type of accident, severity, consequences, type of operations and country of operations (Cummings *et al.*, 2003; Fayad *et al.*, 2003).

In the World Offshore Accident Database (www.dnv.com) for the period 1997–2007, a total of 6033 accidents were registered for the whole world offshore operations. Within that, accidents accounted for 39.7%, incidents and near misses 49.4%, and 10% were insignificant. Table 14.3 displays results from incident data collected on the Norwegian continental shelf where a total of 1223 cases registered for personal injuries covering the period from 2 January 1997 to 29 April 2009 (PSA, 2009). The data on the severity of incidents is based on the PSA criteria showing the decrease in time of incidents with high potential for major accidents or mortalities. A similar decline is seen for incidents with severe consequences.

Table 14.4 shows that consequences and severity of injuries vary based on the type of installation: 1.4% were incidents with a high potential for major accidents; 14.6% were classified as severe incidents; 63.9% of the incidents reported required some follow-up from management and operators.

The data on severity and type of operations was analysed by main activities and classifications. Out of 1223 cases, 405 were not classified by activities, accounting for 33.1% of the database. A total of 358 cases were registered as 'other' and 'not classified', representing 29.27% of total cases registered.

The data displayed in Table 14.5 was analysed by correspondence analysis (Greenacre, 1993) and the result of a symmetric plot of rows and columns is shown in Figure 14.2. From this analysis we can see that drilling well accidents were characterized as severe and that lifting operations incidents were either

Table 14.3   Incidents registered per severity and year.

	Not subject to reporting	Easier follow-up	Potential with small changes	Severe	High potential for major accident/death	Not registered	Total
1997	1	1				28	30
1998		4	3			31	38
1999					1	55	56
2000		1			1	69	71
2001		51		37	3	3	94
2002	1	45		24	6	9	85
2003	2	57	11	10	4		84
2004	4	91	2	26			123
2005	7	100	2	14	1		124
2006	9	117		27			153
2007	4	118		20	1		143
2008	4	155		16			175
up to April 2009		42		5			47
Total	32	782	18	179	17	195	1223

Table 14.4   Incidents registered per severity and type of installation.

Installation type	Not subject to reporting	Some follow-up	Potential with small changes	Severe	High potential for major accident/ death	Not registered	Total frequency	Percent-age
Fixed	9	477	12	94	9	91	692	56.6
Movable	6	140	6	64	6	93	315	25.8
Onshore facility	17	160		19	1		197	16.1
Pipeline				1			1	0.1
Unknown		5		1	1	11	18	1.5
Total	32	782	18	179	17	195	1223	100
%	2.6	63.9	1.5	14.6	1.4	15.9		

not registered or classified as major accidents. For more on correspondence analysis see Chapter 8. For other ways to analyse such data, see Kenett and Raphaeli (2008).

Occupational injury is characterized in this study by four indices. The first index indicates the category, the second indicates the frequency of the injuries, the third refers to the severity of the injury and the fourth refers to actual consequences based on DFU (Defined Hazard For Accidents) classifications. The term

Table 14.5   Incidents registered per severity and type of operations.

	Not subject to reporting	Easier follow-up	Potential with small changes	Severe	High potential for major accident/death	Not registered	Total
Pipeline systems	1	1					2
Lifting operations	1	38	1	28	5	6	79
Constructions and		11		6			17
Compression		3		1			4
Living area	2	41		2			45
Not registered	8	149	3	57	2	186	405
Main process		25	1	3			29
Auxiliary and support systems	4	87	3	15			109
Helicopter transport		2		1			3
Electrics		7	1	3			11
Diving	2	6			1	3	12
Drilling well		101	4	38	6		149
Others	14	311	5	25	3		358
Total					17	195	1223

'DFU' is a Norwegian abbreviation for *Definerte fare- og ulykkessituasjoner*, but is often used also in English, and it refers to defined situations of hazard and accident.

Data related to DFUs with actual consequences is based on the PSA database which relies on data collected in cooperation with the operating companies. These DFUs have been classified, and each type has been given a classification number.

Table 14.6 presents a total of 1223 incidents in the database and 1138 of these were classified as personal injuries, 149 cases as severe accidents and 16 as having a high potential for major accidents and death.

Injury rate per million work hours is calculated as

$$\frac{\text{Number of injuries or accidents}}{\text{Working hours}} \times 1\,000\,000$$

The data from PSA shows that the total injury rate is declining for all activities and types of installations. For mobile and movable installations, the total injury rate declined from 32.8 in 1999 to 8.7 in 2008. And for permanently placed installations, the total injury rate declined from 26.5 in 1999 to 10.7 in 2008.

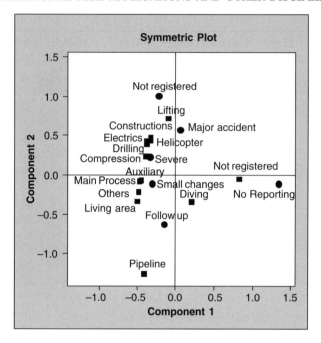

Figure 14.2   Incidents registered per severity and type of operation.

Cost and burden of occupational injuries are considered to be vast and take a big share of industry expenses and national social insurance funds. The European Agency for Safety and Health at Work has estimated that 4.6 million occupational accidents happen every year in the EU, resulting in 146 million lost working hours (EU OSHA, 2001). In the EU, the burden of accidents and injuries is the fourth major cause of death, killing more than 235 000 EU citizens each year.

Occupational accidents are a huge problem in the world; there were over 350 000 fatalities in 2001 (Hämäläinen *et al.*, 2006; Hämäläinen, 2009). Accidents and injuries account for almost 7 million hospital admissions and for 20% of sick leave days. They are the leading cause of death in children and young people. Occupational risk factors are responsible for 8.8% of the global burden of mortality and 8.1% of the combined burden of both mortality and morbidity due to injuries (Concha-Barrientos *et al.*, 2005; Rikhardsson, 2004; Berentsen and Holmboe, 2004; Béjean and Sultan-Taïeb, 2005). Researchers stated that occupational accidents lead to a loss of 2.6 to 3.8% of the collective EU gross national product, every year. Understanding the burden of occupational accidents and incidents is therefore important for setting priorities for prevention and research in OpR (Reville *et al.*, 2001).

Most of these estimates account for only a fraction of the workers' total losses and estimation methods for total accident costs remain controversial. Using a 'willingness to pay approach' based on compensation wage differentials, yields estimates of the order of $6 million (Weil, 2001).

Table 14.6   Incidents registered per severity and DFU.

Actual consequences	Not subject to reporting	Some follow-up	Potential with small changes	Severe	High potential for major accident/death	Not registered	Total
DFU01 leak, non- ignited				1			1
DFU02 leak, ignited					1		1
DFU04 Fire		1					1
DFU10 Personal injury	32	732	15	149	16	194	1138
DFU11 Work-related illness				1			1
DFU14 Radioactive sources		1		1			2
DFU15 Falling objects		20	2	16		1	39
DFU17 Lifting incidents		11		9			20
DFU18 Diving incidents		1					1
DFU36 Others		15		2			17
DFU99 No consequences		1					1
Not registered			1				1
Total	32	782	18	179	17	195	1223

The timing of when injuries occur and when illnesses/diseases are diagnosed, and the allocation of these to a period of analysis, is one of the first issues to be resolved by researchers (NOHSAC, 2006). The time dimension can play an important factor in selecting the appropriateness of the methods used in assessing the economic consequences of occupational accidents, injuries and related sicknesses. Some consequences are immediate as death, injury, disability at the time of an accident, while others can take a longer time to be experienced and are indicated as exposures to hazardous materials in workplaces.

In the oil and gas industry the frequency of severe accidents is limited and therefore does not generate sufficient data for learning opportunities or knowledge sharing. To increase the amount of data, one may augment databases with information from less severe incidents, such as near misses and deviations from established procedures. Such information of near misses and undesirable events can give a relatively good picture of where accidents might occur, but they do not necessarily give a good basis for quantifying risk.

For quantifying risks, hazards and hazardous situations should be identified as individual basic elements contributing to the risk. Risk factors, by themselves, do not necessarily imply a direct impact on safety. There may be other elements contributing to increasing hazards and having a negative impact; these consequences can have an impact on safety, health, environment, business, finance and reputation. These elements were addressed in Chapters 3, 8, 10 and 11.

In order to minimize uncertainties, data reporting should be prioritized and data collection should cover all available sources. When the data is collected from different sources, it is important to ensure that the concepts, definitions, coverage and classifications used by the different sources are consistent. The challenge of merging different databases was the topic of Chapter 5.

The case study in this section was designed to demonstrate the generality of the methods and techniques presented in this book. The final section in the book discusses the challenges and opportunities in operational risks in the area of data management, modelling and decision making.

14.6   Operational risks: data management, modelling and decision making

From the enterprise's point of view, the process of OpR can be divided into six main steps, as schematized in Figure 14.3.

The first two steps consist of collecting and structuring data. The sources of the data may be both internal and external. Under risk assessment, the qualitative component of the value chain in a company's operations and activities is first assessed against a menu of potential operational risk vulnerabilities. This process often incorporates checklists and/or workshops to identify the strengths and weaknesses of the risk management environment. Scenario analysis refers to the phase where companies identify not only past risk events, but also potential risk events based on, for example, past events which happened in other companies of the same industry and the impact of changes in environments on their operations flows. Companies estimate the frequency and severity of these events, identified by analysing the causes of these events and factors causing losses and expanding loss amounts. External data contributes to the development of strong scenario analysis, since it lets the organization rely on experience gained by others in its industry without the organization itself having suffered the damage. For more on risk assessment, see Chapters 2 and 7.

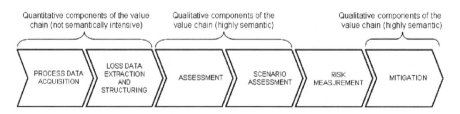

Figure 14.3   The operational risk management process.

Risk measurement is the phase where companies assess the tolerability or acceptability of a risk through estimation of both the probability of its occurrence and the magnitude of its consequence.

Risk mitigation involves prioritizing, evaluating and implementing the appropriate risk-reducing controls recommended from the previous phases. Because the elimination of all risks is usually impractical or close to impossible, it is the responsibility of senior management, and functional and business managers, to use the least cost approach and implement the most appropriate controls to decrease mission risk to an acceptable level, with minimal adverse impact on the organization's resources and mission.

Economic and statistical aspects of risk and OpR have been addressed by several authors (Knight, 1921; Bedford and Cooke, 2001; Tapiero, 2004, 2009, 2010; McNeil *et al.*, 2006; Beaudouin and Munir, 2009; Haimes, 2009). Models derived from physics are recently being applied to analyse stock market behaviour using complex models of system dynamics in a new domain called econophysics (Schinckus, 2009; Shapira *et al.*, 2009, Kenett *et al.*, 2010). This combination of approaches and modelling techniques enhances our ability to conduct effective and efficient risk management activities.

The management of operational risks also has implications for the general topics of safety management, business ethics and social corporate responsibility (Carroll, 1991, 2000; Crane and Matten, 2004; Waldman *et al.*, 2007; Harms-Ringdahl, 2004; Heller, 2006; Whittingham, 2008).

Finally, OpR is also tightly linked to the general topic of asset integrity management (OGP, 2008). The objectives of strategic asset integrity management are:

1. Controlling the asset safely.

2. Achieving high effectiveness in the production process.

3. Providing for effective use of human resources.

4. Extending the lifetime of the asset.

5. Keeping core business functions and controls within the asset.

Rahim *et al.* (2010) propose a new model for achieving all these goals. It is based on five major pillars: competence, compliance, coordination, communication and control. This 5C model has been applied to the oil and gas industry with substantial success.

The above section illustrates the wide scope of OpR covering technical, mathematical, statistical, psychological, semantic, economical and sociological areas. We conclude the chapter and book with a final summary section.

14.7 Summary

This concluding chapter complements the landscape of modern Operational risk management (OpR) presented in this book. As was shown, OpR is a continuously developing area of active research and applications involving a wide range of disciplines. Many practitioners, managers in general and risk managers in particular fully realize that operational risks are a source of significant potential losses and that they should be better understood and better managed. With the rising complexity of business operations, and with higher demands and expectations from the customers, operations can no longer be delegated to a lower echelon in the organization and, certainly, cannot be assumed to be risk-free.

Consequently, many organizations dedicate more attention to proactive assessment and mitigation of such risks. The material presented in the book refers to novel approaches in OpR, including those developed in the MUSING project (MUSING, 2006). This involves the application of semantic analysis capabilities and state-of-the-art statistical models in analysing operational risk data. The book also looks at the analysis of near misses and opportunity losses, the merging of OpR-related data sources, new association rules for text mining in OpR textual logs and applications of Bayesian networks for mapping causality links.

The concepts presented throughout the book are shown, in this final chapter, to be relevant to areas of business activity such as the energy industry and air travel, not just to the financial sector regulated by Basel II. Moreover, the chapter discusses the convergence of quality management and risk management and the role of 'black swans'. Both disciplines aim at satisfying various stakeholders' demands such as compliance, efficiency, sustainability and business continuity by weighing unforeseen events against their possible effect on the 'bottom line'. To quote Taleb:

> One needs a lot of skills to get a BMW, a lot of skills + a lot of luck to get a private jet.
>
> N. Taleb (Personal communication, 2010)

Like any organizational activity, the domain of OpR requires management involvement. Effective planning, deployment and improvement of OpR require management with maturity at the top of the quality ladder (Section 14.3). The book provides methods, techniques and examples for managers who want and are able to handle risks proactively and to go up the quality ladder. After all, a risk, once identified, is no longer a risk – it is a management problem.

References

Amalberti, R. (2001) The paradoxes of almost totally safe transportation systems. *Safety Science*, 37, pp, 109–126.

Andersson, R. and Menckel, E. (1995) On the prevention of accidents and injuries – a comparative analysis of conceptual frameworks. *Accident Analysis and Prevention*, 27, pp. 757–768.

Ashby, W.R. (1958) Requisite variety and its implications for the control of complex systems. *Cybernetica*, 1, 2, pp. 83–99.

Attwood, D., Khan, F. and Veitch, B. (2006) Occupational accident models – where have we been and where are we going? *Journal of Loss Prevention in the Process Industries*, 19, 6, pp. 664–682.

Aven, T. (2003) *Foundation of Risk Analysis: A Knowledge and Decision-Oriented Perspective*, John Wiley & Sons, Ltd, Chichester.

Aven, T. and Vinnem, J.E. (2007) *Risk Management with Applications from the Offshore Petroleum Industry, Springer Series in Reliability Engineering*, Springer-Verlag, Berlin.

Barach, P. and Small, D.S. (2000) Reporting and preventing medical mishaps: lessons from non-medical near miss reporting systems. *British Medical Journal*, 320, pp. 759–763.

Bea, R.G. (2001) Risk assessment and management of offshore structures. *Progress in Structural and Engineering Materials*, 3, pp. 180–187.

Beaudouin, F. and Munir, B. (2009) A revision of industrial risk managements: decisions and experimental tools in risk business. *Risk and Decision Analysis*, 1, pp. 3–20.

Bedford, T. and Cooke, R. (2001) *Probabilistic Risk Analysis: Foundations and Methods*, Cambridge University Press, Cambridge.

Béjean, S. and Sultan-Taïeb, H. (2005) Modelling the economic burden of disease imputable to stress at work. *European Journal of Health Economics*, 50, pp. 16–23.

Berentsen, R. and Holmboe, R.H. (2004) Incidents/accidents classification and reporting in Statoil. *Journal of Hazardous Materials*, 111, pp. 155–159.

Carroll, A.B. (1991) The pyramid of corporate social responsibility: toward the moral management of organizational stakeholders. *Business Horizons*, 34, 4, pp. 39–48.

Carroll, A.B. (2000) The four faces of corporate citizenship, in *Business Ethics*, Richardson, J.E. (Ed.), McGraw-Hill, Guilford, CT.

Chichilnisky, G. (2010) The foundations of statistics with black swans. *Mathematical Social Sciences*, 59, 2, pp. 184–192.

Concha-Barrientos, M., Nelson, D.I., Fingerhut, M., Driscoll, T. and Leigh, J. (2005) The global burden due to occupational injury. *American Journal of Industrial Medicine*, 48, pp. 470–481.

Crane, A. and Matten, D. (2004) *Business Ethics: A European Perspective – Managing Corporate Citizenship and Sustainability in the Age of Globalization*, Oxford University Press, Oxford.

CSB – US Chemical Safety Board (2007) *US Chemical Safety and Hazard Investigation Board*. The report of the BP US Refineries independent safety review panel.

Cummings, P., McKnight, B. and Greenland, S. (2003) Matched cohort methods for injury research. *Epidemiologic Reviews*, 25, pp. 43–50.

EU OSHA – European Agency for Safety and Health at Work (2001) *Economic Impact of Occupational Safety and Health in the Member States of the European Union*, EU European Agency for Safety and Health at Work, Bilbao.

Fayad, R., Nuwayhid, I., Tamim, H., Kassak, K. and Khogali, M. (2003) Cost of work-related injuries in insured workplaces in Lebanon. *Bulletin of the World Health Organization*, 81, pp. 509–516.

Godfrey, A.B. and Kenett, R.S. (2007) Joseph M. Juran: a perspective on past contributions and future impact. *Quality and Reliability Engineering International*, 23, pp. 653–663.

Greenacre, M.J. (1993) *Correspondence Analysis in Practice*, Harcourt, Brace and Company: Academic Press, London.

Haimes, Y.Y. (2009) *Risk Modelling, Assessment and Management*, 3rd Edition, John Wiley & Sons, Inc., Hoboken, NJ.

Hämäläinen, P. (2009) The effect of globalization on occupational accidents. *Safety Science*, 47, pp. 733–742.

Hämäläinen, P., Takala, J. and Saarela, K.L. (2006) Global estimates of occupational accidents. *Safety Science*, 44, pp. 137–156.

Harms-Ringdahl, L. (2004) Relationships between accident investigations, risk analysis, and safety management. *Journal of Hazardous Materials*, 111, pp. 13–19.

Heller, S. (2006) Managing industrial risk – having a tested and proven system to prevent and assess risk. *Journal of Hazardous Materials*, 130, pp. 58–63.

Johnsen, J. and Everson, C. (2003) *Preventing Accidents*, 4th Edition, Institute of Leadership and Management, London.

Jones, S., Kirchsteiger, C. and Bjerke, W. (1999) The importance of near miss reporting to further improve safety performance. *Journal of Loss Prevention in the Process Industries*, 12, 1, pp. 59–67.

Juran, J.M. (1986) The Quality Trilogy: a universal approach to managing for quality. *Proceedings of the ASQC 40th Annual Quality Congress*, Anaheim, California.

Juran, J.M. (1989) *Juran on Leadership for Quality – An Executive Handbook*, Free Press, New York.

Kenett, R.S. (2007) Cause and effect diagrams, in F. Ruggeri, R.S. Kenett and F. Faltin (Eds), *Encyclopaedia of Statistics in Quality and Reliability*, John Wiley & Sons, Ltd, Chichester.

Kenett, R.S. (2008) From data to information to knowledge. *Six Sigma Forum Magazine*, pp. 32–33.

Kenett, R.S. (2009) Discussion of post-financial meltdown: what do the services industries need from us now? *Applied Stochastic Models in Business and Industry*, 25, pp. 527–531.

Kenett, R.S. and Baker, E. (2010) *Process Improvement and CMMI for Systems and Software: Planning, Implementation, and Management*, Auerbach Publications, Boca Raton, FL.

Kenett, R.S. and Kenett, D.A. (2008) Quality by design applications in biosimilar technological products. *ACQUAL, Accreditation and Quality Assurance*, 13, 12, pp. 681–690.

Kenett, R.S. and Raphaeli, O. (2008) Multivariate methods in enterprise system implementation, risk management and change management. *International Journal of Risk Assessment and Management*, 9, 3, pp. 258–276.

Kenett, R.S. and Salini, S. (2008a) Relative linkage disequilibrium applications to aircraft accidents and operational risks. *Transactions on Machine Learning and Data Mining*, 1, 2, pp. 83–96.

Kenett, R.S. and Salini, S. (2008b) Relative linkage disequilibrium: a new measure for association rules, in P. Perner (Ed.), *Advances in Data Mining: Medial Applications, E-Commerce, Marketing, and Theoretical Aspects*, ICDM 2008, Lecture Notes in Computer Science Vol. 5077, Springer-Verlag, Berlin.

Kenett, R.S. and Shmueli, G. (2009) On Information quality. University of Maryland, School of Business Working Paper RHS 06-100, http://ssrn.com/abstract=1464444 (accessed 21 May 2010).

Kenett, R.S. and Tapiero, C. (2010) Quality, risk and the Taleb quadrants. *Risk and Decision Analysis*, 4, 1, pp. 231–246.

Kenett, R.S. and Zacks, S. (1998) *Modern Industrial Statistics: Design and Control of Quality and Reliability*, Duxbury Press, San Francisco.

Kenett, R.S., de Frenne, A., Tort-Martorell, X. and McCollin, C. (2008) The statistical efficiency conjecture, in S. Coleman *et al.* (Eds), *Statistical Practice in Business and Industry*, John Wiley & Sons, Ltd, Chichester.

Kenett, D.Y., Shapira, Y. and Ben-Jacob, E. (2010) RMT assessments of the market latent information embedded in the stocks' raw, normalized, and partial correlations. *Journal of Probability and Statistics*, DOI: 10.1155/2009/249370.

Knight, F.H. (1921) *Risk, Uncertainty and Profit*, Houghton Mifflin, Boston, MA, 1964.

Kogan, K. and Tapiero, C.S. (2007) *Supply Chain Games: Operations Management and Risk Valuation*, Series in Operations Research and Management Science, Springer-Verlag, Berlin.

Mandelbrot, B. (1982) *The Fractal Geometry of Nature*, W H Freeman, San Francisco.

McKinnon, R. (2000) *Cause, Effect, and Control of Accidental Loss with Accident Investigation Kit*, Lewis, London.

McNeil, A.J., Frey, R. and Embrechts, P. (2006) *Quantitative Risk Management: Concepts, Techniques, and Tools*, Princeton University Press, Princeton, NJ.

Meeker, W.Q. and Escobar, L. (2004) Reliability: the other dimension of quality. *Quality Technology and Quantitative Management*, 1, 1, pp. 1–25.

Muermann, A. and Oktem, U. (2002) The near miss management of operational risk. *Journal of Financial Risk*, Fall, pp. 25–36.

MUSING (2006) IST-FP6 27097, http://www.musing.eu (accessed 21 May 2010).

NASA – National Aeronautics and Space Administration (2002) *Probabilistic Risk Assessment Procedures Guide for NASA Managers and Practitioners*, Version 1.1, Office of Safety and Mission Assurance, Washington, DC.

Nasr, M. (2007) Quality by Design (QbD) – a modern system approach to pharmaceutical development and manufacturing – FDA perspective. FDA *Quality Initiatives Workshop*, Washington, DC.

Nazeri, Z., Barbara, D., De Jong, K., Donohue, G. and Sherry, L. (2008) *Contrast-Set Mining of Aircraft Accident and Incident Data*, Technical Report, George Mason University.

NOHSAC – National Occupational Health and Safety Advisory Committee (2006) *Access Economics: The economic and social costs of occupational disease and injury in New Zealand*, NOHSAC Technical Report 4, Wellington.

OGP – International Association of Oil and Gas Producers (2008) *Asset integrity – the key to managing major incident risks*. Report No. 415.

Phimister, J.R., Oktem, U., Kleindorfer, P.R. and Kunreuther, H. (2003) Near miss incident management in the chemical process industry. *Risk Analysis*, 23, pp. 445–459.

PSA – Petroleum Safety Authority Norway (2009) *Safety status and signals 2008-2009*, www.ptil.no (accessed 21 May 2010).

Raanan, Y. (2009) Risk management in higher education - do we need it? *Sinergie*, 78, pp. 43–56.

Rahim, Y. (2010) Accident, incident reporting systems and investigations: consequences, costs and burden on safety management systems in oil and gas industry. PhD Thesis,

Department of Statistics and Applied Mathematics, Faculty of Economics University of Turin.

Rahim, Y. and Kenett, R.S. (2008a) Factors affecting the near miss reporting. *European Network for Business and Industrial Statistics (ENBIS) Eighth Annual Conference*, Athens, Greece.

Rahim, Y. and Kenett, R.S. (2008b) Using risk assessment matrices in ranking projects and modifications in the petroleum industry. *Modeling and Analysis of Safety and Risk in Complex Systems*, MASR 2008, St Petersburg, Russia.

Rahim, Y. and Kenett, R.S. (2009) Challenges in estimating the cost and burden of occupation accidents. *European Network for Business and Industrial Statistics (ENBIS) Ninth Annual Conference*, Goteborg, Sweden.

Rahim, Y., Refsdal, I. and Kenett, R.S. (2010) The 5C model: a new approach to asset integrity management. *International Journal of Pressure Vessels and Piping*, 87, 3, pp. 88–93

Reville, R.T., Bhattacharya, J. and Weinstein, L.S. (2001) New methods and data sources for measuring economic consequences of workplace injuries. *American Journal of Industrial Medicine*, 40, pp. 452–463.

Rikhardsson, P.M. (2004) Accounting for the cost of occupational accidents corporate social responsibility. *Environmental Management*, 11, pp. 63–70.

Ruggeri, F., Kenett, R.S. and Faltin, F. (2007) *Encyclopaedia of Statistics in Quality and Reliability*, 4 Vols, John Wiley & Sons, Ltd, Chichester.

Schinckus, C. (2009) Economic uncertainty and econophysics. *Physica A*, 388, pp. 4415–4423.

Schofield, S. (1998) Offshore QRA and the ALARP principle. *Reliability Engineering and System Safety*, 61, pp. 31–37.

Shapira, Y., Kenett, D.Y. and Ben-Jacob, E. (2009) The index cohesive effect on stock market correlations. *European Journal of Physics B*, 72, 4, pp. 657–669.

Swuste, P. (2007) Qualitative methods for occupational risk prevention strategies in safety or control banding. *Safety Science Monitor*, 1, 3, Article 8.

Taleb, N.N. (2007) *The Black Swan: The impact of the highly improbable*, Random House, New York.

Taleb, N.N. (2008a) The fourth quadrant: a map of the limits of statistics. *Edge*, http://www.edge.org/3rd_culture/taleb08/taleb08_index.html (accessed 21 May 2010).

Taleb, N.N. (2008b) Errors, robustness and the fourth quadrant. Working paper, The Center for Risk Engineering, NYU-Polytechnic Institute.

Tapiero, C. (2004) *Risk and Financial Management: Mathematical and Computational Methods*, John Wiley & Sons, Hoboken, NJ.

Tapiero, C. (2009) The price of safety and economic reliability. NYU-Poly Technical Report, www.ssrn.com/abstract=1433477 (accessed 21 May 2010).

Tapiero, C. (2010) The price of quality claims. *Applied Stochastic Models in Business and Industry*, www.ssrn.com/abstract=1433498 (accessed 21 May 2010).

van der Schaaf, T. and Kanse, L. (2004) Biases in incident reporting databases: an empirical study in the chemical process industry. *Safety Science*, 42, 1, pp. 57–67.

Waldman, D., Kenett, R.S. and Zilberg, T. (2007) Corporate social responsibility: what it really is, why it's so important, and how it should be managed (in Hebrew). *Status Magazine*, 193, pp. 10–14.

Weil, D. (2001) Valuing the economic consequences of work injury and illness: a comparison of methods and findings. *American Journal of Industrial Medicine*, 40, pp. 418–437.

Whittingham, R.B. (2008) *Preventing Corporate Accidents: An Ethical Approach*. Elsevier, Oxford.

Yashchin, E. (2007) Modelling of risk losses using size-biased data. *IBM Journal of Research & Development*, 51, 3/4, pp. 309–323.

Index

Operational Risk Management: A Practical Approach to Intelligent Data Analysis   Edited by Ron S. Kenett
and Yossi Raanan   © 2011 John Wiley & Sons, Ltd

Statistics in Practice

Human and Biological Sciences

Earth and Environmental Sciences

Buck, Cavanagh and Litton – Bayesian Approach to Interpreting Archaeological Data
Glasbey and Horgan – Image Analysis in the Biological Sciences
Helsel – Nondetects and Data Analysis: Statistics for Censored Environmental Data
Illian, Penttinen, Stoyan, H and Stoyan D – Statistical Analysis and Modelling of Spatial Point Patterns
McBride – Using Statistical Methods for Water Quality Management
Webster and Oliver – Geostatistics for Environmental Scientists, Second Edition
Wymer (Ed) – Statistical Framework for Recreational Water Quality Criteria and Monitoring

Industry, Commerce and Finance

Aitken – Statistics and the Evaluation of Evidence for Forensic Scientists, Second Edition
Balding – Weight-of-evidence for Forensic DNA Profiles
Brandimarte – Numerical Methods in Finance and Economics: A MATLAB-Based Introduction, Second Edition
Brandimarte and Zotteri – Introduction to Distribution Logistics
Chan – Simulation Techniques in Financial Risk Management
Coleman, Greenfield, Stewardson and Montgomery (Eds) – Statistical Practice in Business and Industry
Frisen (Ed) – Financial Surveillance
Fung and Hu – Statistical DNA Forensics
Gusti Ngurah Agung – Time Series Data Analysis Using EViews
Jank and Shmueli (Ed.) – Statistical Methods in e-Commerce Research
Lehtonen and Pahkinen – Practical Methods for Design and Analysis of Complex Surveys, Second Edition
Ohser and Mücklich – Statistical Analysis of Microstructures in Materials Science
Pourret, Naim & Marcot (Eds) – Bayesian Networks: A Practical Guide to Applications
Taroni, Aitken, Garbolino and Biedermann – Bayesian Networks and Probabilistic Inference in Forensic Science
Taroni, Bozza, Biedermann, Garbolino and Aitken – Data Analysis in Forensic Science